Big Ideas in Mathematics
for Future Mathematics Teachers

Big Ideas in Modern Geometry

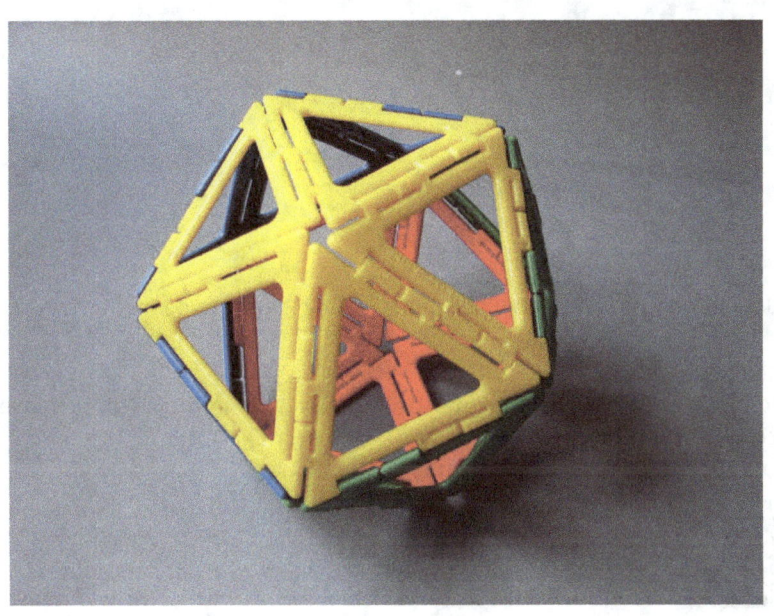

John Beam, Jason Belnap, Eric Kuennen,
Amy Parrott, and Jennifer Szydlik
(Updated Summer 2024)

Copyright 2024 by John Beam, Jason Belnap, Eric Kuennen, Amy Parrott and Jennifer Szydlik

This work is licensed under the Creative Commons Attribution-NonCommercial-NoDerivatives 4.0 International License. To view a copy of this license, visit http://creativecommons.org/licenses/by-nc-nd/4.0/ or send a letter to Creative Commons, PO Box 1866, Mountain View, CA 94042, USA.

Cover of photo credit: Geocentric armillary sphere-MHS 1344, Musée d'histoire des sciences de la Ville de Genève, by Rama, Public domain, via Wikimedia Commons

Title page photo credit: Nevit Dilmen, CC BY-SA 3.0, via Wikimedia Commons

First Edition 2021

ISBN: 9798526840798

Eric W. Kuennen, Mathematics Department, University of Wisconsin Oshkosh
800 Algoma Blvd, Oshkosh, WI 54901

Dear Future Teacher,

We wrote this book to help you to see the structure that underlies geometry, to give you experiences really *doing* mathematics, and to show you how children think and learn. We fully intend this course to transform your relationship with math.

As teachers of future math teachers, we created or gathered the activities for this text, and then we tried them out with our own students and modified them based on their suggestions and insights. We know that some of the problems are tough – you will get stuck sometimes. Please don't let that discourage you. There's much value in wrestling with an idea.

All our best,

John, Jason, Eric, Amy, & Jen

Table of Contents

TABLE OF CONTENTS	3
HEY! READ THIS. IT WILL HELP YOU UNDERSTAND THE BOOK.	6
COMMON CORE STATE STANDARDS FOR MIDDLE GRADES GEOMETRY	8
CORE STATE STANDARDS FOR MATHEMATICAL PRACTICE	9
COMMON CORE STATE STANDARDS FOR MATHEMATICAL PRACTICE	10
CLASS ACTIVITY 1: TOWN RULES	14
AXIOMATIC SYSTEMS AND MODELS	15
CLASS ACTIVITY 2: TWO FINITE GEOMETRIES	20
AFFINE AND PROJECTIVE FINITE GEOMETRY AXIOMS, PARALLELISM	22
USING PRECISE MATHEMATICAL LANGUAGE	23
CLASS ACTIVITY 3: POINTS OF PAPPUS	29
CLASS ACTIVITY 4: READING EUCLID	30
EUCLID'S AXIOMS	32
EUCLIDEAN LINES AND ANGLES	34
CLASS ACTIVITY 5: ENOUGH IS ENOUGH	37
POLYGONS AND TRIANGLES	38
CONGRUENCE	39
TRIANGLE CONGRUENCE THEOREMS	40
CLASS ACTIVITY 6: CONSTRUCTION ZONE	46
STRAIGHT EDGE AND COMPASS CONSTRUCTION	47
CLASS ACTIVITY 7: IF YOU BUILD IT	52
SUMMARY OF BIG IDEAS FROM CHAPTER ONE	54
CLASS ACTIVITY 8: CIRCULAR REASONING	56
INCENTER, ORTHOCENTER, CIRCUMCENTER, AND CENTROID	56
EUCLIDEAN CIRCLES	57
VAN HIELE LEVELS	60
CLASS ACTIVITY 9: FINDING FORMULAS	65
LENGTH AND AREA	67
MAKING SENSE OF AREA FORMULAS	68

CLASS ACTIVITY 10: PLAYING PYTHAGORAS — 70
The Pythagorean Theorem — 71

CLASS ACTIVITY 11: NOTHING BUT NET — 75
Right Versus Oblique Prisms — 75
Polyhedra — 76
Volume and Surface Area — 78

CLASS ACTIVITY 12: SLIDES, TURNS AND FLIPS — 81
Rigid Motions of the Plane (Isometries) — 84

CLASS ACTIVITY 13: FLIP-FLOP — 91
Compositions of Isometries — 91

CLASS ACTIVITY 14: TRANSFORMATIVE THINKING — 100
Transformation and Isometry Groups — 100

CLASS ACTIVITY 15: TESSELLATIONS — 104
Tessellations — 106
Regular Tessellations — 106

CLASS ACTIVITY 16: EXPANDING AND CONTRACTING — 110

CLASS ACTIVITY 17: STRICTLY PLATONIC (SOLIDS) — 115
Symmetries in Space — 117

CLASS ACTIVITY 18: BURIED TREASURE — 122
The Cartesian Plane — 123
Analytic Geometry — 125

CLASS ACTIVITY 19: PLAGUE OF LOCUS — 130
The Conic Sections — 133

SUMMARY OF BIG IDEAS FROM CHAPTER TWO — 136

CLASS ACTIVITY 20: LIFE ON A ONE-SIDED WORLD — 138
The Möbius Strip — 139
ion Spaces — 140

CLASS ACTIVITY 21: LIFE IN A TAXICAB WORLD — 143
Measuring Distance — 144
Circles and Triangles — 145

CLASS ACTIVITY 22: LIFE ON A SPHERICAL WORLD — 148
Lines and Distance on Spheres — 150
Parallelism Triangles on Spheres — 151

CLASS ACTIVITY 23: LIFE ON A HYPERBOLIC WORLD — 157
Parallel Lines in Hyperbolic Space — 159

CLASS ACTIVITY 24: LIFE IN A FRACTAL WORLD — 162
Self-similarity: Natural and Mathematical — 164
Iterative Process and Dimension — 166

SUMMARY OF BIG IDEAS FROM CHAPTER THREE — 171

REFERENCES: — 173

EUCLID'S POSTULATES AND PROPOSITIONS — 174
Book I — 174
Book I Propositions — 175

GLOSSARY — 180

POLYGON CUT-OUTS (FOR READ AND STUDY 10) — 191

HYPERBOLIC PAPER TEMPLATE — 197

Hey! Read this. It will help you understand the book.

The only way to learn mathematics is to do mathematics.
Paul Halmos

This book was written to prepare future teachers for the mathematical work of teaching. The focus of this module is geometry, and mathematics doesn't get any better than that. Geometry allows us to think spatially, to see structure in art and form, and to create and visualize new "worlds" with different rules.

Does the word "geometry" call to mind the *two-column-proof* of your high school days? Long ago mathematics educators decided that geometry class would be a good place to showcase the importance of definitions, reasoning, and proof in mathematical thinking – really, these things are vital in *all areas* of mathematics – not just geometry – and if you use any of our other modules, you'll see that this is so. However, if the two-column proof has ruined geometry for you, then forget about it. You don't need to do any here. You are free to reason in any form you see fit, as long as you can communicate your argument to others. After all, we mathematicians rarely write a proof in such a form. *We'd* hate to be constrained in that way.

Geometry is a domain for action and activities. The National Council of Teachers of Mathematics (NCTM) advocates that middle grades students draw, measure, visualize, compare, transform, and classify geometric objects (NCTM, 2000). (Note all the action verbs.) We will do all these things in this module. The ideas in this book are fundamentally important for your students to understand and so they are fundamentally important for *you* to understand.

As mathematicians, we will also try to convey to you the beauty of our subject. Mathematicians view mathematics as the study of patterns and structures. We want to show you how to reason like a mathematician – and we want you to show this to your students too. This *way of reasoning* is just as important as any content you teach. When you stand before your class, you are a representative of the mathematical community; we will help you to become a good one.

No one can do this thinking for you. Mathematics isn't a subject you can memorize; it is about ways of thinking and knowing. *You* need to do examples, gather data, look for patterns, experiment, draw pictures, think, try again, make arguments, and think some more. The big ideas of probability and statistics are not always easy.

Each section of this book begins with a **Class Activity**. The activity is designed for small-group work in class. Some activities may take your class as little as 30 minutes to complete and discuss. Others may take you two or more class periods. No solutions are provided to activities – you will have to solve them yourselves. The **Read and Study**, **Connections to Teaching**, and **Homework** sections are presented within the context of the activity ideas.

> *One of the big misapprehensions about mathematics that we perpetrate in our classrooms is that the teacher always seems to know the answer to any problem that is discussed. This gives students the idea that there is a book somewhere with all the right answers to all of the interesting questions, and that teachers know those answers. And if one could get hold of the book, one would have everything settled. That's so unlike the true nature of mathematics.*
>
> *Leon Henkin*

To prepare ourselves to write this text we studied four *Standards*-based curriculum projects for middle school students (the books your future students might use). Those projects are *Mathematics in Context*, *Connected Mathematics*, *MATHematics*, and *MathScape*. All of these are activity-based curricula. This means that the middle school materials were written so that your future students will solve problems and create understandings based on concrete experiences.

In case you are skeptical about these types of materials for your future students, let us assure you that they better encourage and support the types of behaviors and thinking that mathematicians value than do traditional materials. Furthermore, the research suggests that schools that had adopted *Standards*-based materials for more than two years showed significantly higher test scores on even traditional measures of mathematical understanding than did matched schools that adopted traditional curricula (Reys, Reys, Lapan, Holliday, & Wasman, 2003; Riordan & Noyce, 2001; Griffen, Evans, Timms, & Trowell, 2000). We assure you that the ideas you will meet in these pages are vitally connected to the mathematics curriculum of your future students, and we hope that the text is written in a way that makes these connections apparent to you.

The Common Core State Standards is "a state-led effort to establish a shared set of clear educational standards for English language arts and mathematics that states can voluntarily adopt. The standards have been informed by the best available evidence and the highest state standards across the country and globe and designed by a diverse group of teachers, experts, parents, and school administrators..." (see http://www.corestandards.org/Math/) As of the time of publication of this text, most states had officially adopted these standards, and so it is important for you to know them and the content and practices that they advocate.

The mathematics content in this book is focus on preparing you to teach the Common Core State Standards for Mathematics for grades 6 - 8. These are the standards that you will likely follow when you are a teacher, so we will highlight aspects of them throughout the text. In order for you to see how the mathematical work you are doing appears in the elementary grades, we have made explicit connections to *Core Connections* from College Preparatory Mathematics (CPM). This is the middle grades mathematics curriculum adopted by the Oshkosh Area School District. You will often be asked to go to the site https://cpm.org/university to read or do problems. Your instructor will provide you with a code so that you can access these materials.

Common Core State Standards for Middle Grades Geometry

Grade 6
Solve real-world and mathematical problems involving area, surface area, and volume.

1. Find the area of right triangles, other triangles, special quadrilaterals, and polygons by composing into rectangles or decomposing into triangles and other shapes; apply these techniques in the context of solving real-world and mathematical problems.

2. Find the volume of a right rectangular prism with fractional edge lengths by packing it with unit cubes of the appropriate unit fraction edge lengths, and show that the volume is the same as would be found by multiplying the edge lengths of the prism. Apply the formulas $V = lwh$ and $V = bh$ to find volumes of right rectangular prisms with fractional edge lengths in the context of solving real-world and mathematical problems.

3. Draw polygons in the coordinate plane given coordinates for the vertices; use coordinates to find the length of a side joining points with the same first coordinate or the same second coordinate. Apply these techniques in the context of solving real-world and mathematical problems.

4. Represent three-dimensional figures using nets made up of rectangles and triangles, and use the nets to find the surface area of these figures. Apply these techniques in the context of solving real-world and mathematical problems.

Grade 7
Draw, construct, and describe geometrical figures and describe the relationships between them.

1. Solve problems involving scale drawings of geometric figures, including computing actual lengths and areas from a scale drawing and reproducing a scale drawing at a different scale.

2. Draw (freehand, with ruler and protractor, and with technology) geometric shapes with given conditions. Focus on constructing triangles from three measures of angles or sides, noticing when the conditions determine a unique triangle, more than one triangle, or no triangle.

3. Describe the two-dimensional figures that result from slicing three-dimensional figures, as in plane sections of right rectangular prisms and right rectangular pyramids.

Solve real-life and mathematical problems involving angle measure, area, surface area, and volume.

4. Know the formulas for the area and circumference of a circle and use them to solve problems; give an informal derivation of the relationship between the circumference and area of a circle.

5. Use facts about supplementary, complementary, vertical, and adjacent angles in a multi-step problem to write and solve simple equations for an unknown angle in a figure.

6. Solve real-world and mathematical problems involving area, volume and surface area of two- and three-dimensional objects composed of triangles, quadrilaterals, polygons, cubes, and right prisms.

Copyright 2010. National Governors Association Center for Best Practices and Council of Chief State School Officers. All rights reserved.

Grade 8

Understand congruence and similarity using physical models, transparencies, or geometry software.

1. Verify experimentally the properties of rotations, reflections, and translations:

    a Lines are taken to lines, and line segments to line segments of the same length.

    b Angles are taken to angles of the same measure.

    c Parallel lines are taken to parallel lines.

2. Understand that a two-dimensional figure is congruent to another if the second can be obtained from the first by a sequence of rotations, reflections, and translations; given two congruent figures, describe a sequence that exhibits the congruence between them.

3. Describe the effect of dilations, translations, rotations, and reflections on two-dimensional figures using coordinates.

4. Understand that a two-dimensional figure is similar to another if the second can be obtained from the first by a sequence of rotations, reflections, translations, and dilations; given two similar two-dimensional figures, describe a sequence that exhibits the similarity between them.

5. Use informal arguments to establish facts about the angle sum and exterior angle of triangles, about the angles created when parallel lines are cut by a transversal, and the angle-angle criterion for similarity of triangles. *For example, arrange three copies of the same triangle so that the sum of the three angles appears to form a line, and give an argument in terms of transversals why this is so.*

Understand and apply the Pythagorean Theorem.

6. Explain a proof of the Pythagorean Theorem and its converse.

7. Apply the Pythagorean Theorem to determine unknown side lengths in right triangles in real-world and mathematical problems in two and three dimensions.

8. Apply the Pythagorean Theorem to find the distance between two points in a coordinate system.

Solve real-world and mathematical problems involving volume of cylinders, cones, and spheres.

9. Know the formulas for the volumes of cones, cylinders, and spheres and use them to solve real-world and mathematical problems.

Copyright 2010. National Governors Association Center for Best Practices and Council of Chief State School Officers. All rights reserved.

Common Core State Standards for Mathematical Practice

Mathematics | Standards for Mathematical Practice

The Standards for Mathematical Practice describe varieties of expertise that mathematics educators at all levels should seek to develop in their students. These practices rest on important "processes and proficiencies" with longstanding importance in mathematics education. The first of these are the NCTM process standards of problem solving, reasoning and proof, communication, representation, and connections. The second are the strands of mathematical proficiency specified in the National Research Council's report *Adding It Up*: adaptive reasoning, strategic competence, conceptual understanding (comprehension of mathematical concepts, operations and relations), procedural fluency (skill in carrying out procedures flexibly, accurately, efficiently and appropriately), and productive disposition (habitual inclination to see mathematics as sensible, useful, and worthwhile, coupled with a belief in diligence and one's own efficacy).

1 Make sense of problems and persevere in solving them.

Mathematically proficient students start by explaining to themselves the meaning of a problem and looking for entry points to its solution. They analyze givens, constraints, relationships, and goals. They make conjectures about the form and meaning of the solution and plan a solution pathway rather than simply jumping into a solution attempt. They consider analogous problems, and try special cases and simpler forms of the original problem in order to gain insight into its solution. They monitor and evaluate their progress and change course if necessary. Older students might, depending on the context of the problem, transform algebraic expressions or change the viewing window on their graphing calculator to get the information they need. Mathematically proficient students can explain correspondences between equations, verbal descriptions, tables, and graphs or draw diagrams of important features and relationships, graph data, and search for regularity or trends. Younger students might rely on using concrete objects or pictures to help conceptualize and solve a problem. Mathematically proficient students check their answers to problems using a different method, and they continually ask themselves, "Does this make sense?" They can understand the approaches of others to solving complex problems and identify correspondences between different approaches.

2 Reason abstractly and quantitatively.

Mathematically proficient students make sense of quantities and their relationships in problem situations. They bring two complementary abilities to bear on problems involving quantitative relationships: the ability to *decontextualize*—to abstract a given situation and represent it symbolically and manipulate the representing symbols as if they have a life of their own, without necessarily attending to their referents—and the ability to *contextualize*, to pause as needed during the manipulation process in order to probe into the referents for the symbols involved. Quantitative reasoning entails habits of creating a coherent representation of the problem at hand; considering the units involved; attending to the meaning of quantities, not just how to compute them; and knowing and flexibly using different properties of operations and objects.

3 Construct viable arguments and critique the reasoning of others.

Mathematically proficient students understand and use stated assumptions, definitions, and previously established results in constructing arguments. They make conjectures and build a logical progression of statements to explore the truth of their conjectures. They are able to analyze situations by breaking them into cases, and can recognize and use counterexamples. They justify their conclusions,

Copyright 2010. National Governors Association Center for Best Practices and Council of Chief State School Officers. All rights reserved.

communicate them to others, and respond to the arguments of others. They reason inductively about data, making plausible arguments that take into account the context from which the data arose. Mathematically proficient students are also able to compare the effectiveness of two plausible arguments, distinguish correct logic or reasoning from that which is flawed, and—if there is a flaw in an argument—explain what it is. Elementary students can construct arguments using concrete referents such as objects, drawings, diagrams, and actions. Such arguments can make sense and be correct, even though they are not generalized or made formal until later grades. Later, students learn to determine domains to which an argument applies. Students at all grades can listen or read the arguments of others, decide whether they make sense, and ask useful questions to clarify or improve the arguments.

4 Model with mathematics.
Mathematically proficient students can apply the mathematics they know to solve problems arising in everyday life, society, and the workplace. In early grades, this might be as simple as writing an addition equation to describe a situation. In middle grades, a student might apply proportional reasoning to plan a school event or analyze a problem in the community. By high school, a student might use geometry to solve a design problem or use a function to describe how one quantity of interest depends on another. Mathematically proficient students who can apply what they know are comfortable making assumptions and approximations to simplify a complicated situation, realizing that these may need revision later. They are able to identify important quantities in a practical situation and map their relationships using such tools as diagrams, two-way tables, graphs, flowcharts and formulas. They can analyze those relationships mathematically to draw conclusions. They routinely interpret their mathematical results in the context of the situation and reflect on whether the results make sense, possibly improving the model if it has not served its purpose.

5 Use appropriate tools strategically.
Mathematically proficient students consider the available tools when solving a mathematical problem. These tools might include pencil and paper, concrete models, a ruler, a protractor, a calculator, a spreadsheet, a computer algebra system, a statistical package, or dynamic geometry software. Proficient students are sufficiently familiar with tools appropriate for their grade or course to make sound decisions about when each of these tools might be helpful, recognizing both the insight to be gained and their limitations. For example, mathematically proficient high school students analyze graphs of functions and solutions generated using a graphing calculator. They detect possible errors by strategically using estimation and other mathematical knowledge. When making mathematical models, they know that technology can enable them to visualize the results of varying assumptions, explore consequences, and compare predictions with data. Mathematically proficient students at various grade levels are able to identify relevant external mathematical resources, such as digital content located on a website, and use them to pose or solve problems. They are able to use technological tools to explore and deepen their understanding of concepts.

6 Attend to precision.
Mathematically proficient students try to communicate precisely to others. They try to use clear definitions in discussion with others and in their own reasoning. They state the meaning of the symbols they choose, including using the equal sign consistently and appropriately. They are careful about specifying units of measure, and labeling axes to clarify the correspondence with quantities in a problem. They calculate accurately and efficiently, express numerical answers with a degree of precision appropriate for the problem context. In the elementary grades, students give carefully formulated explanations to each other. By the time they reach high school they have learned to examine claims and make explicit use of definitions.

Copyright 2010. National Governors Association Center for Best Practices and Council of Chief State School Officers. All rights reserved.

7 Look for and make use of structure.

Mathematically proficient students look closely to discern a pattern or structure. Young students, for example, might notice that three and seven more is the same amount as seven and three more, or they may sort a collection of shapes according to how many sides the shapes have. Later, students will see 7×8 equals the well remembered $7 \times 5 + 7 \times 3$, in preparation for learning about the distributive property. In the expression $x^2 + 9x + 14$, older students can see the 14 as 2×7 and the 9 as $2 + 7$. They recognize the significance of an existing line in a geometric figure and can use the strategy of drawing an auxiliary line for solving problems. They also can step back for an overview and shift perspective. They can see complicated things, such as some algebraic expressions, as single objects or as being composed of several objects. For example, they can see $5 - 3(x - y)^2$ as 5 minus a positive number times a square and use that to realize that its value cannot be more than 5 for any real numbers x and y.

8 Look for and express regularity in repeated reasoning.

Mathematically proficient students notice if calculations are repeated, and look both for general methods and for shortcuts. Upper elementary students might notice when dividing 25 by 11 that they are repeating the same calculations over and over again, and conclude they have a repeating decimal. By paying attention to the calculation of slope as they repeatedly check whether points are on the line through (1, 2) with slope 3, middle school students might abstract the equation $(y - 2)/(x - 1) = 3$. Noticing the regularity in the way terms cancel when expanding $(x - 1)(x + 1)$, $(x - 1)(x^2 + x + 1)$, and $(x - 1)(x^3 + x^2 + x + 1)$ might lead them to the general formula for the sum of a geometric series. As they work to solve a problem, mathematically proficient students maintain oversight of the process, while attending to the details. They continually evaluate the reasonableness of their intermediate results.

Connecting the Standards for Mathematical Practice to the Standards for Mathematical Content

The Standards for Mathematical Practice describe ways in which developing student practitioners of the discipline of mathematics increasingly ought to engage with the subject matter as they grow in mathematical maturity and expertise throughout the elementary, middle and high school years. Designers of curricula, assessments, and professional development should all attend to the need to connect the mathematical practices to mathematical content in mathematics instruction.

The Standards for Mathematical Content are a balanced combination of procedure and understanding. Expectations that begin with the word "understand" are often especially good opportunities to connect the practices to the content. Students who lack understanding of a topic may rely on procedures too heavily. Without a flexible base from which to work, they may be less likely to consider analogous problems, represent problems coherently, justify conclusions, apply the mathematics to practical situations, use technology mindfully to work with the mathematics, explain the mathematics accurately to other students, step back for an overview, or deviate from a known procedure to find a shortcut. In short, a lack of understanding effectively prevents a student from engaging in the mathematical practices.

In this respect, those content standards which set an expectation of understanding are potential "points of intersection" between the Standards for Mathematical Content and the Standards for Mathematical Practice. These points of intersection are intended to be weighted toward central and generative concepts in the school mathematics curriculum that most merit the time, resources, innovative energies, and focus necessary to qualitatively improve the curriculum, instruction, assessment, professional development, and student achievement in mathematics.

Copyright 2010. National Governors Association Center for Best Practices and Council of Chief State School Officers. All rights reserved.

Chapter One

Arguing from Axioms

Class Activity 1: Town Rules

The mathematician starts with a few propositions, the proof of which is so obvious that they are called self-evident. The rest of (his) work consists of subtle deductions from them.
                                                                                              Thomas Henry Huxley

Welcome to the super fun town of Hilbert! We have a few rules here just to be sure all our residents have plenty of friends and hobbies. In Hilbert, a **club** is a membership list, and no two distinct clubs have the same membership list. Here are the rules legislated for our clubs:

a) Every two townspeople have a club to which they both belong, and that club is unique (meaning that for each pair of people there is only one such club).

b) Every club has at least two members.

c) No club contains all the townspeople.

d) If you name a club and a townsperson who is not a member of that club, there will be one and only one club that person belongs to that has no members in common with the first club.

We are interested in how many people could live in Hilbert and follow these rules. Check for town populations of one through five. In each case, argue that you are correct. Any conjectures (guesses) about town populations larger than five?

Read and Study

Geometry is the science of correct reasoning on incorrect figures.
George Polya

What is geometry all about? During this course we will try to give you several different ways of thinking about geometry. The first is this:

Geometry is the study of ideal shapes and spaces and the relationships that exist among them.

The word *ideal* is important. Geometry – and all mathematics for that matter – is not about real objects. *Think about it, have you ever seen a circle?* You've seen plenty of *representations* of circles, but an actual circle (all points in a plane equidistant from a given point) exists only in our minds. And so do points and planes. All mathematical objects are like this – they are ideas.

Doing geometry (and all mathematics) in a formal sense means starting with some defined (and some undefined) ideal objects, and some rules (called axioms) and reasoning to see what is "true" about the objects. We put "true" in quotes, because we don't mean true in a theological sense, but rather we mean "true" within the system we have created.

Hilbert is an example of an **axiomatic system**. We described an object called a "club," we gave you some rules (axioms) about the behavior of these clubs, and then we left you figure out what was "true" about the town.

Hilbert is also a geometry. Change the rules about townspeople and clubs to rules about points and lines and you will see what we mean. *Take a minute to compare the following rules to the ones in the Town Rules activity.* (Did you notice the italics? That is your signal to do something. Mathematicians read math books with pencil in hand. We answer questions and verify anything the authors claim to be true. Start doing this too. The italics will help you remember to *slow down and think while you read*.)

a) Every two points are on a unique line.
b) Every line contains at least two points.
c) No line contains all the points.
d) If you name a line and a point not on the line, there will be one and only one line on the point that is parallel to the given line.

Axiomatic systems include five parts: undefined terms (like 'member'), defined terms (like 'club'), axioms (like "Every club has at least two members."), theorems (things you can deduce from the axioms, like 'Hilbert cannot have a population of exactly two people.') and proofs of theorems (arguments that the theorems are true based on the axioms).

Okay, before we go any further, let's clarify some of the language that mathematicians use to talk about the process of doing mathematics. Here are some important definitions:

1) **axiom**: a rule that the mathematical community has decided to accept as true without proof. An axiom is an *assumption*.

2) **conjecture**: a conjecture is a hypothesis or a guess about what is true given the axioms. For example, after some experience with Hilbert, you might have *conjectured* that the number of people in Hilbert must be a perfect square.

3) **inductive reasoning**: coming to a conclusion based on examples. I might notice that the sun rose the day before yesterday, it rose yesterday, and it rose today; so I might conclude that the sun will rise tomorrow. This is inductive reasoning. This type of reasoning is often used to generate a conjecture, but it is *not* considered sufficient evidence by mathematicians to prove a general statement.

4) **deductive reasoning**: coming to conclusion based on the axioms and logic. This type of reasoning is the hallmark of mathematical argument.

5) **counterexample**: a counterexample is a specific example that shows that a conjecture is false. *Give an example of a counterexample.*

6) **proof**: a mathematical proof consists of a deductive argument that establishes the truth of a claim.

7) **theorem**: a theorem is a mathematical statement that has been proven to be true. For example, it is a theorem that Hilbert cannot have a population of exactly three people. This is not stated as a specific axiom, but you can deduce this based on the axioms. Our argument goes something like this:

    Suppose Abe, Ben, and Cal live in Hilbert. Then, because rule (a) says that each pair must belong to a unique club together, we must have Club 1 consisting of, say, Abe and Ben, Club 2 consisting of Abe and Cal, and Club 3 consisting of Ben and Cal. *Make certain you understand why we must have these three clubs when we follow rule (a).*

    We cannot have 3 people in any club because rule (c) states that all of the townspeople cannot belong to one club together. And we cannot have any clubs of only one person because rule (b) says that each club must have at least two members. So Clubs 1, 2, and 3 are the only possible clubs we can make and we must have each of them to follow rule (a). *Make certain you understand why rules (b) and (c) force us to conclude that Clubs 1, 2, and 3 are the only possible clubs we can make.*

Thus we have only one case (Clubs 1, 2, and 3) to investigate with respect to rule (d). Rule (d) says that if I name a club (say, Club 2) and a person not in that club (say, Ben), I must find one and only one other club to which Ben belongs that does not contain Abe or Cal. But that is not possible since there are only two other clubs, Club 1 which contains Abe and Club 3 which contains Cal. *Make certain you understand why we cannot follow rule (d).*

Therefore, since we have already argued that we cannot form any other clubs, there is no way Ben can belong to a club without Abe or Cal. Thus rule (d) cannot be met with exactly three people living in Hilbert and we have made our argument it is not possible for exactly three people to live in Hilbert.

Take some time here and ask yourself if you really understand what you just read. Did you answer all of the questions? Can you explain the argument to someone else? Most students are in the habit of reading textbooks too casually. The previous section is tough – it could easily take a careful reader 20 to 30 minutes to read the preceding four paragraphs with understanding. Remember what we said about mathematicians reading slowly and thoughtfully with a pencil in hand? Taking the time to read this text like a mathematician would is one of the surest ways to deepen your understanding of geometry. (Another way is to do all the homework problems with the same type of careful thought.)

There are a couple of things we strive for when creating an axiomatic system. First, we need that the axioms be **consistent**. In other words, the axioms shouldn't contradict one another. Second, we want the system as lean as possible – no **redundancy**.

If (and only if) the axioms are consistent, then there exists a **model** for the system. (Kurt Gödel proved this theorem in 1930.) In making the argument above we attempted to create a model of the system for three people. We gave specific names to the three people (Abe, Ben, and Cal) and names and member lists to the three clubs we formed: Club 1 = {Abe, Ben}, Club 2 = {Abe, Cal}, and Club 3 = {Ben, Cal}. To form a model we simply identify each object in the system with a concrete representation in such a way that the definitions and rules of the system make sense. A model is a kind of "super example." It is a concrete way to "see" a mathematical structure.

There are several possible models for the first three rules of the People and Clubs system for three people. We already saw that the people could be named and the clubs could be modeled as sets of names. For another model, we could let each person be one of the letters **A, B, C** and let a club be represented by a line segment. This model would look like this:

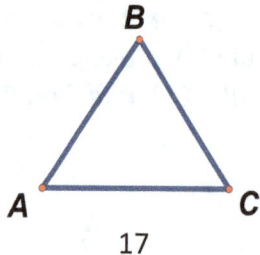

Even though these two models *look* different they represent the same system, the same set of information about the relationships between the three people (points) and the clubs (lines) to which they belong. Sometimes it is very useful to have more than one way to look at, or represent, the same thing in mathematics. It is important to recognize when different-looking objects have the same underlying structure or the same set of properties.

You probably found a four-person model for the town of Hilbert. Maybe you thought of a set of names and club membership lists that satisfied all Hilbert's axioms. Perhaps you drew a set of points and lines to show clubs that met all the axioms. In any case, you created a set of objects and interpretations for the undefined terms in such a manner that all the axioms were true at the same time using your interpretations. That is, you created a model for the axiom system. If the axioms are not consistent, there can be no model. *Read this whole section again. It is important to understand these ideas*.

And now we can give you a second definition of geometry. A **geometry** is an axiomatic system about objects called "points" and collections of points called "lines" and the relationships between points and lines, that is we say a point is "on" a line and a line is "on" (or contains) a point.

It may surprise you to learn that there many geometries. Euclidean geometry is the one taught in school, and it is very useful for working on flat surfaces. However, if we want to talk about lines and triangles on a curved surface such as a sphere (also a practical concern, since we live on the surface of a sphere), we need a different geometry. The rules of Euclid can't all be obeyed on a sphere. The Hilbert system for four people is also a geometry.

Before we move on to make some connections to the middle grades, there are several important words used in the Town Rules that we need to talk about. The words are *every, unique, at least, one and only one,* and *all*. These words are examples of what mathematicians call **quantifiers**, words that tell us something important about how many objects are involved in the statement. Other quantifying words and phrases are *some, exactly, at most, each, there is*. It is crucial to understand the distinctions between these words and to use them carefully in making arguments. You will be asked to do so in the homework set.

Homework:

Children are not vessels to be filled, but lamps to be lighted.
Henry Steele Commager

1. Go back and do all the things in italics in the *Read and Study* section.

2. Consider the following six statements. Which carry the same meaning? Which can be true at the same time, even though they do not carry the same meaning? Why? Which cannot be true at the same time? Why?
   a) There is a cat living in my house.
   b) There are three cats living in my house.
   c) I have exactly one dog living in my house.
   d) There is one and only one dog living in my house.
   e) There are at least three animals living in my house.
   f) Some of the animals living in my house have four legs.

3. Underline each of the quantifiers found in the statements in the preceding problem. Explain what each one tells you about the number of animals living in my house

4. Prove that $1 + 2 + 3 + \ldots + (n-1) + n = \frac{1}{2}[n \times (n+1)]$.

5. Here is an axiomatic system with the undefined terms *corner*, *square*, and *on* and the following axioms:

   I. There is a square.
   II. Each square is on exactly four distinct corners.
   III. For each square, there are exactly four distinct squares with exactly two corners on the given square.
   IV. Each corner is on exactly four distinct squares.

   a) Create an infinite model *in the plane* for this system.
   b) Create a finite model for just the first three axioms (in other words, your set of objects will be finite).
   c) See if you can create a finite model for all four axioms.

6. Here are axioms for Triad Geometry:

   I. There are exactly three points.
   II. Each pair of points is on exactly one line.
   III. No line contains all the points.

   a) Make a model for this finite geometry using dots as points and segments as lines. Can there be more than one configuration that satisfies these axioms? Explain.
   b) Now make a model using letters as points and pairs of letters as lines.
   c) Your first model with the dots and segments was a **Euclidean model** (one based on geometry in the infinite flat plane) and it might have been misleading because *line segments in the Euclidean sense do not exist in finite geometry*. List some other familiar Euclidean objects that don't exist in any finite geometry.

Class Activity 2: Two Finite Geometries

Projective geometry is all geometry.

Arthur Cayley

Here is an axiomatic system for **Affine Plane Finite Geometries**:

We have a finite set of 'points' and 'lines' so that the following are true (note that again 'point,' 'line' and 'on' are undefined terms):

    I.   Every two distinct points have exactly one line on them both.
    II.   Given a line and a point not on that line, there is exactly one line on the point that has no points on the first line.
    III.   Every line is on at least two points.
    IV.   There exist three non-collinear points.

An affine plane with *n* points on each line is said to have **order *n***.

a) Sketch a model for an affine plane of order 2.
b) Is Hilbert an affine geometry? Explain.
c) Here is model for an affine plane of order 3. Check to see that it satisfies all the axioms.

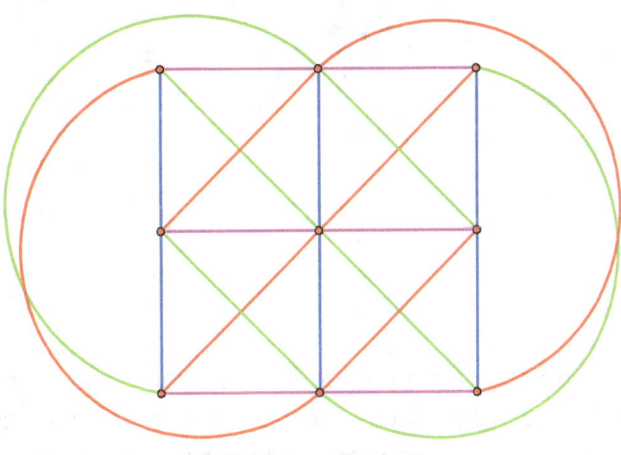

Affine plane of order 3

(This activity is continued on the next page.)

Here is an axiomatic system for **Projective Plane Finite Geometries**.

    I.   Every two distinct points have exactly one line on them both.
    II.   Every two lines have exactly one point on them both.
    III.   Every line is on at least three points.
    IV.   There exist three non-collinear points.

A projective plane of **order n** has $n + 1$ points on each line.

We are going to describe how to sketch a model of a projective plane of order two by starting with a model of the affine plane of order two (see below) and adding some structure. Here is the plan for your group to follow: Collect the lines parallel to each other in a class (in our picture, each set of mutually parallel lines is the same color). For each of the $n + 1$ classes of parallel lines (in this case there are three classes), add a new point that will be on each of those lines (you may need to extend them and curve them around so that they intersect). Then add another new line containing exactly all of these $n + 1$ new points. Try it.

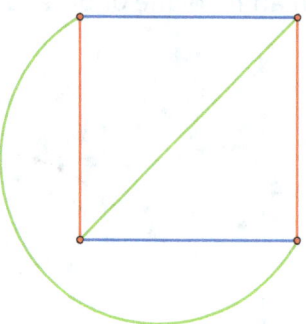

    d) How many points are on each line? How many lines on each point? How many points total are in a projective plane of order 2? How many lines?
    e) Given a line and a point not on the line, how many lines are there through the given point that are parallel to the given line?

Read and Study

> *... to characterize the import of pure geometry, we might use the standard form of a movie-disclaimer: No portrayal of the characteristics of geometrical figures or of the spatial properties of relationships of actual bodies is intended, and any similarities between the primitive concepts and their customary geometrical connotations are purely coincidental.*
>
> Carl G. Hempel, in The World of Mathematics

A **finite geometry** consists of a finite number of objects (typically called points) and their relationships (typically described in terms of points being 'on' lines). The terms 'point,' 'line,' and the relationship 'on' are usually undefined. The properties are established by a set of axioms that govern the relationships. Hilbert is an example of a finite geometry if we think of people as points and clubs as lines and the town rules as the axioms. Triad Geometry is another example of a finite geometry.

There are two types of finite geometries that are of particular interest to mathematicians: affine and projective. (In fact some mathematicians define a finite geometry in such a way that these are the *only* two types of finite geometries.) The model you made in the class activity is an example of our first projective geometry. What is the primary distinction between these two flavors of finite geometry? In affine plane geometry, through a point not on a given line, we get *one* parallel line. In projective plane geometry, we get *none*.

Look at the axioms for each geometry to see which axiom tells you about parallelism. It might help to recall that two lines *k* and *l* are **parallel**, written *k* || *l*, if they are in the same plane and no point is on both *k* and *l*.

Mathematicians have proven that affine planes of order n exist whenever $n = p^k$ (where p is prime and k is a whole number). But we don't know yet which other values of n give us affine planes. We call that an *open question*. People are probably working on it right now.

In the class activity, you worked to understand axioms and create models for the axiomatic systems. We know that working with axioms isn't easy. As you work on this text, please keep in mind that our aim is to help you understand what a geometry is from the perspective of mathematicians. The specific geometries we introduce and the specific theorems about them are not as important as the idea that there are many geometries, each with its own axioms and its own models, and there is a way of thinking and a language that we use to study them all.

This *Read and Study* is devoted to the language of mathematics that you will need to study axiom systems. In particular, below we'll discuss several distinctions that are not always important in everyday speech but are vital for understanding mathematical arguments.

Let's do it.

Distinction 1: (A statement and its *negation*). If we have some statement P, then the negation of P is the statement "Not P." If P is true, then "not P" is false. And if P is false, then "not P" is true. For example, the negation of the statement, "I love geometry," is the statement "I do not love geometry."

Distinction 2: (*or* versus *and*.) Here is a case when math-speak differs a bit from everyday conversation. When a mathematician says something like 'x is an element of A **or** B,' (Here assume A and B are sets – notice how we use capitals to denote sets and small letters for elements of sets – this is pretty typical – but not a rule or anything) she means that x could be in A, x could be in B or **x could be in *both* at the same time.**

When she says "x is an element of A **and** B" she means x is definitely in both. *Decide whether each of the following statements is true:*

1) A square has four sides and a triangle has four sides.
2) A square has four sides or a triangle has four sides.
3) A square has four sides and a triangle has three sides.
4) A square has four sides or a triangle has three sides

Now let's see how we negate an "and" sentence like 3) A square has four sides **and** a triangle has three sides.

Notice that for the above sentence to be true (which it is), *both* parts must be true. If the above sentence is not true then either a square doesn't have four sides or a triangle doesn't have three sides. That means the negation is the sentence:

A square does not have four sides **or** a triangle does not have three sides.

In other words, "not (A **and** B)" means "(not A) **or** (not B)." *Stop and think this through. Write the negation of the sentence: I ate peas and potatoes.*

What about the negation of an "or" sentence? Consider (true) sentence 2): A square has four sides **or** a triangle has four sides.

For this to be false, both a square can't have four sides and a triangle can't have four sides. So "not (P **or** Q)" is equivalent to "(not P) **and** (not Q)."

23

A square does not have four sides **and** a triangle does not have four sides.

Write the negation of the sentence: x is an element of set A or x is an element of set B.

Distinction 3: (*converse* versus *contrapositive*) Many mathematical statements are conditional statements ("if-then" statements). Here are some examples. *Decide whether each is true or false, and in each case explain your thinking.*

1) If a polygon is a square, then it is a rectangle.

2) If a polygon is a rectangle, then it is a square.

3) If you live in Los Angeles, then you live in California.

4) If you don't live in California, then you don't live in Los Angeles.

5) If it is Friday, then tomorrow is Saturday.

Mathematicians call the statement 'If Q, then P' the **converse** of 'If P, then Q'. So Statement 2 is the converse of Statement 1 above. Note that those two statements are not **logically equivalent** (not always true at the same time.).

A statement of the form 'If not Q, then not P' is called the **contrapositive** of the statement 'If P, then Q.' The contrapositive form *is* logically equivalent to 'If P, then Q". So Statement 3 and Statement 4 above are logically equivalent statements. *Think about it to make sure that seems right. What is the converse of Statement 5 above? How about the contrapositive?*

Sometimes when you want to prove that an if/then statement is true, it is easier to prove that the contrapositive statement is true than it is to prove the original statement is true. And the point here is that by proving the contrapositive you also prove the original (because they are equivalent statements).

Distinction 4: (*"There exists..."* versus *"For all..."*) Recall that these sentence starters are called quantifiers, and they tell you whether the statement is claiming that something exists (there is at least one), or whether the statement is a general one (meaning that it is true for every case). Here are some examples:

1. Every club has at least four members.
2. There exists a club with exactly four members.
3. For every club and for each townsperson who is not a member of that club, there will exist one and only one club that person belongs to that has no members in common with the first club.

Notice that the first statement makes a claim about *all* clubs, whereas the second statement only makes a claim about at least *one* club. The third statement uses a combination of quantifiers – which brings us to our next distinction.

Distinction 5: ('*There exists an x, such that for all y...*' versus '*For all y, there exists an x, such that...*') Below are two statements that mean exactly the same thing in real life talk, but have quite different meanings in mathematics. *Can you figure out how the following statements might be different?*

There is someone for everyone.

For everyone, there is someone.

In mathematics-speak, the first statement says that there is one person for the entire group -- one person who is for *all* of us. The second statement says that each of us has our own special person. For each of us, there is someone, and *my* someone may be different from yours (at least I hope so).

Here are some examples of how this looks in a mathematical context. *Decide whether each is true or false. Make an argument in each case.* For now, assume both x and y must be integers (elements of the set {...-3, -2, -1, 0, 1, 2, 3 ...}).

a) For all x, there exists a y, such that $x + y = 0$.
b) There exists an x, such that for all y, $x + y = 0$.
c) There exists an x, such that for all y, $xy = 0$.
d) For all x and for all y, $x + y$ is an integer.
e) For all x and for all y, $x + y = 7$.
f) There exists an x and there exists a y, such that $x + y = 7$.
g) For all x, there exists a y such that $x + y = 7$.

Connections to Teaching:

Mathematically proficient students understand and use stated assumptions, definitions, and previously established results in constructing arguments.

You may have noticed that we have been talking quite a bit about facets of "doing mathematics" in general in these first sections. This is because *your* mathematical practices – your habits of mind and habits of behavior regarding math – will provide a context for all of your thinking and work in geometry, and they will also give your future students a model of what it means to do mathematics.

Building on the work of the National Council of Teachers of Mathematics (NCTM), the team of educators and mathematicians who wrote the Common Core Standards for Mathematics singled out eight Standards for Mathematical Practice that students should learn during the course of their schooling. It will be your job, as their teacher, to help them to establish these mathematics practices. *Carefully read the Common Core Standards for Mathematical Practice near the beginning of this textbook.* We will return to these throughout the text.

Homework:

Discovery consists of seeing what everybody has seen and thinking what nobody has thought.

Albert Szent-Gyorgyi

1) If you haven't already done so, go back and do all the things in italics in the *Read and Study* section.

2) The first Common Core Standard for Mathematical Practice is about making sense of problems and persevering in solving them. Read that standard again. To what extent do you monitor your own thinking as you solve a problem? Did you do any of the things described as you worked on the *Class Activity*? Explain.

3) Explain, as you would to your middle grades students, why a statement of the form 'not (A or B)' is equivalent to '(not A) and (not B).' An example may help.

4) Here is a theorem about the town of Hilbert: If a club in Hilbert has exactly *n* members, then all of the clubs have exactly *n* members.

    a) State the converse of this theorem. Is it true?

    b) State the contrapositive of this theorem. Is it true?

5) Eric's Geometry has the following undefined terms: book, library, on; and this set of axioms:

    Axiom I: There is at least one book.
    Axiom II: Each library has exactly four books on it.
    Axiom III: Each book has exactly two libraries on it.

    a) Make a model that satisfies the axioms.
    b) Use your model to make some conjectures about Eric's Geometry.
    c) See if you can prove that one of your conjectures is true.
    d) Is Eric's Geometry a finite geometry? Explain.
    e) Write the negation of Axiom III.

6) The third Common Core Standard for Mathematical Practices is about making arguments. Read that paragraph again. Describe some specific things that you did when you worked on the Town Rules *Class Activity* that would fit that standard.

7) Spend 15 minutes on this question: Is it possible to have a finite geometry where if you are given a line and a point not on the line, you can have **more than one line** through the point that is parallel to the given line? Recall that parallel lines in finite geometry need not 'look parallel.' Rely on the *definition* of "parallel" to help you think about this.

8) It turns out that you can always turn affine plane models into projective plane models by doing the modification you did in the class activity: Collect the lines parallel to each other in a class. For each of the $n + 1$ classes of parallel lines, add a new point that will be on each of those lines. Then define all of these $n + 1$ new points to all be on the same line. See if you can create a model for a projective plane of order 3 by modifying the affine plane of order 3 below as described. Then check to be sure your model fulfills all the projective plane axioms.

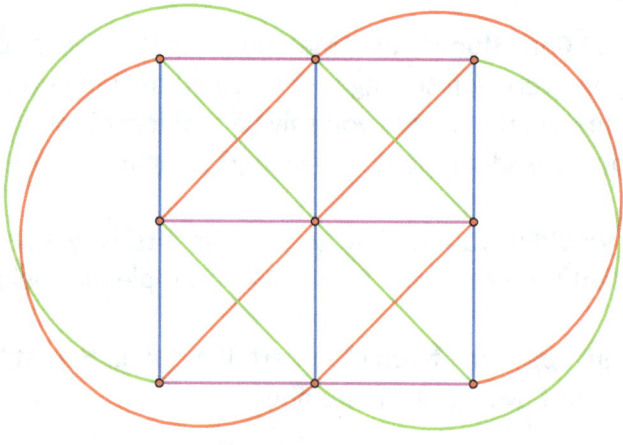

Affine plane of order 3

9) Here are two theorems about affine planes:

   Affine Theorem 1: If some line of an affine plane has n points on it, then *each* line has n points on it and *each* point has $n + 1$ lines on it.

   Affine Theorem 2: In an affine plane if some line has n points on it, then there are n^2 points and $n(n + 1)$ lines, and each line has n lines parallel to it (including itself).

   Here are two theorems about projective planes:

   Projective Theorem 1: If one line of a projective plane has $n + 1$ points on it, then all lines have $n + 1$ points on them and all points have $n + 1$ lines on them.

   Projective Theorem 2: In a projective plane if some line has $n + 1$ points on it, then there are $n^2 + n + 1$ points and $n^2 + n + 1$ lines.

a) Check to see that our model for an affine plane of order 3 satisfies Theorems 1 and 2 above.
b) Check your projective plane of order 2 from the class activity and see if it satisfies both theorems.
c) State the contrapositive of Affine Theorem 1.
d) State the converse of Projective Theorem 1.
e) Compare the theorems about affine planes to the projective planes theorems.
f) Use the affine plane axioms to prove that the minimum number of points in any affine plane is four and the minimum number of lines is six.

Class Activity 3: Points of Pappus

Geometry enlightens the intellect and sets one's mind right.

Ibn Khaldun (MQS)

In the first half of the fourth century Pappus of Alexandria wrote a guide to Greek geometry titled *The Mathematical Collection*. In that guide he discussed the work of Euclid, Archimedes and Ptolemy, presenting their theorems, constructions and arguments.

Carefully read the following theorem of Pappus:

> If A, B and C are three distinct points on one line and A', B' and C' are three different distinct points on a second line, then the intersection of line AC' and line CA', line AB' and BA', and line BC' and CB' are collinear (the three intersection points all lie on the same line).

Under the requirement that the specified lines intersect, this becomes a Euclidean Theorem, meaning that it is true in the familiar flat infinite plane of your high school days.

1) Draw some careful sketches, using different configurations of A, B and C and A', B' and C' and see if this theorem seems to be true. Does it still hold if B' isn't between A' and C'? Just to make it so we can all talk about this as a class, label the intersection point of AB' and BA' as D, the intersection of AC' and CA' as E, and the intersection of BC' and CB' as F. What situations must be avoided to ensure that all nine points exist?

2) Now, let's leave the Euclidean world and consider *just* the nine points of Pappus along with their "lines" as a finite geometry. (In other words, now, *no other points exist except the nine and lines are just sets of points*.)

   a) How many lines appear on your sketches? How many points on each line? How many lines on each point?

   b) Given a line and a point not on the line, how many other lines contain the given point *and* intersect the given line? State a conjecture based on your observations. What sort of counterexample would be required in order to prove your conjecture false?

   c) Given a line and a point not on the line, how many lines on the given point *are not* on the given line? State a conjecture based on your observations. What sort of counterexample would be required in order to prove your conjecture false?

   d) See if you can create the axioms for which this system is a model.

Class Activity 4: Reading Euclid

Euclid taught me that without assumptions there is no proof. Therefore, in any argument examine the assumptions.

Eric Temple Bell in H. Eves Return to Mathematical Circles

In your group, carefully study the **postulates** (another word for *axioms*) of Euclid's Geometry. These are basically the original formulations from Euclid's text – but Euclid wrote in Greek and not in English, so they have been translated for you. Take out your compass (circle maker) and straightedge (line maker) and see how the postulates correspond with these tools.

Euclid's Postulates (Axioms)
(quoted from Thomas L. Heath's translation of *Euclid's Elements*, 2002)

Let the following be postulated:

1. To draw a straight line from any point to any point.

2. To produce a finite straight line continuously in a straight line.

3. To describe a circle with any center and distance.

4. That all right angles are equal to one another.

5. That, if a straight line falling on two straight lines makes the interior angles on the same side less than two right angles, the two straight lines, if produced indefinitely, meet on that side on which are the angles less than the two right angles.

On the next page, you will find the first proof that appears in Euclid's text. Study it.

Proposition 1

On a given finite straight line to construct an equilateral triangle.

Let *AB* be the given finite straight line.

Thus it is required to construct an equilateral triangle on the straight line *AB*.

With centre *A* and distance *AB* let the circle *BCD* be described;   [Post. 3]
again, with centre *B* and distance *BA* let the circle *ACE* be described;   [Post. 3]
and from the point *C*, in which the circles cut one another, to the points *A*, *B* let the straight lines *CA*, *CB* be joined.   [Post. 1]

Now, since the point *A* is the centre of the circle *CDB*,
             AC is equal to *AB*.   [Def. 15]

Again, since the point *B* is the centre of the circle *CAE*,
             BC is equal to *BA*.   [Def. 15]

But *CA* was also proved equal to *AB*;
        therefore each of the straight lines *CA*, *CB* is equal to *AB*.

And things which are equal to the same thing are also equal to one another;   [C.N. 1]
        therefore *CA* is also equal to *CB*.

Therefore the three straight lines *CA*, *AB*, *BC* are equal to one another.

Therefore the triangle *ABC* is equilateral; and it has been constructed on the given finite straight line *AB*.

                                           Being what it was required to do.

Used with permission from Heath, T.L. (2002) translation of *Euclid's Elements*.
D. Densmore (Ed.) Green Lion Press, Santa Fe, New Mexico. pp. 3.

What exactly is Euclid proving here?

What things do you notice about the form of his argument?

Euclid cites Def. 15: A *circle* is a plane figure contained on one line such that all the straight lines falling upon it from one point among those lying within the figure are equal to one another (Heath, 2002, p.1). Does this correspond with definition of a **circle** we provided in the glossary? Explain.

Read and Study:

> *A point is that which has no part.*
> *A line is a breadthless length.*
>
> *Euclid, Elements*

In this chapter we will explore the world of geometry created by Euclid's postulates (axioms). Euclid lived after Plato in Greece around 300 BC. He is known primarily for his work on *Elements*, a text that laid axiomatic foundations for geometry in the plane. This text has had a tremendous influence on mathematics because of the systematic way it presents geometry propositions (theorems) logically derived from one another.

Euclid's geometry is a world of flat planes covered with infinitely many points and no holes, endless straight lines, and circles that look just like the circle of your elementary school days. His is the world where if you see a line and a point not on that line, you will find exactly *one* line parallel to the given line. His is the world in which you did your high school geometry. In fact, in high school his geometry was your *only* geometry; what we want you to know now is that Euclidean geometry is but *one of many* geometries. We have already seen some other geometries that are finite. Later we will study some non-Euclidean geometries that contain infinitely many points.

Here are the **Euclidean Postulates (Axioms)** perhaps written in a more user-friendly form than that which you saw in the Class Activity:

1) A unique straight line segment can be drawn from any point to any other point.

2) A straight line segment can be extended to produce a unique straight line.

3) A circle may be described with any center and distance.

4) All right angles are equal to each other.

5) Version A: If two lines are cut by a transversal and the interior angles on the same side are less than two right angles, then the lines will meet on that side.

   Version B: Through a given point not on a line, there can be drawn only one line parallel to the given line.

   (These two versions of the Fifth Postulate are equivalent – and for the purposes of this course, you can use whichever one is most convenient for you in any given argument. Version B is also known as Playfair's Axiom.)

As you saw in the *Class Activity*, with the exception of number four, these axioms are all *constructive*. By this we mean that they are about what can be constructed using only a straightedge (a 'line maker') and a compass (a 'circle maker'). Axiom four is a little bit different. It tells us that no matter where we are on the plane, all right angles are congruent. So it provides us with the idea that Euclid's plane is uniform in some way – that is, no matter where you raise a perpendicular or drop a perpendicular, the angles you construct will all be the same.

Euclid also listed in the *Elements* some additional axioms (like the below) that he called Common Notions.

1) Things equal to the same thing are also equal to one another.

2) If equals are added to equals, the wholes are equal.

3) If equals are subtracted from equals, the remainders are equal.

4) Things which coincide with one another are equal.

From this lean set of tools, Euclid then carefully began to build the theorems (he called them **propositions**) of his geometry.

It is worth noting here that today's mathematicians have found Euclid's set of axioms a bit *too* lean, and they have added many more axioms to Euclidean geometry. For example, in Proposition 1, when Euclid gave a proof that he could construct an equilateral triangle, he made two circles each having the radius of the given segment and used a point where those circles intersected to identify a vertex of the triangle. Modern mathematicians would note that he was implicitly assuming that those two circles *would* intersect in a point (that point wouldn't be missing from the geometry or anything), and have decided that there should be an axiom to that affect. However, Euclid did a pretty good job overall – and 2300 years later his book *Elements* is still the "bible" of geometry.

We also want to note that Euclid often used the word "equal" when we would use the word "congruent." Today's mathematicians use "equal" when they want to compare two *numbers*. So we might say that ½ is equal to 0.5. We use the word "congruent" when we want to say that two objects (like two triangles or two segments) are the same size and shape. The basic idea here is that two objects are *congruent* in the case where if one object was moved to lie on top of the other object, they would correspond exactly. We will do a more careful job of defining "congruent" later.

Just like we did in our finite geometry worlds, we will now try to see what theorems we can prove using Euclid's assumptions. In fact the game we will play for the rest of the chapter and the next is this: whenever you are asked to prove a Euclidean theorem, you should turn to the Appendix where Euclid's postulates and propositions are listed and find it. Then you are free to use (assume) any postulate and any proposition listed *before* the one you are trying to prove. For example, say you want to prove that in an **isosceles** triangle, the base angles are congruent. *Go to the Appendix (really do it) and see if you can find that theorem. Then come right back here.*

Since that particular theorem is part of Proposition 5, that means you may use any of the Propositions 1 – 4 as well as any of the postulates in your argument, and as you use them, you should cite them.

Before we do an example to show you what an argument might look like, you will need to review some relevant Euclidean Geometry definitions about parallel lines and angles. (In this geometry, point, line, plane and angle are undefined terms.) *Spend some time reviewing the following terms.*

- Two angles are **supplements** if together they make two right angles.

- Two angles are **complements** if together they make a right angle.

- **Vertical angles** are angles opposite each other when two lines intersect in a point.

- Two lines are **parallel** if they lie in the same plane and share no common point.

- Two lines are **perpendicular** if they form right vertical angles at a point of intersection.

- Two objects are **congruent** if they can be made to coincide with one another. (If you moved one on top of the other, it would fit exactly.)

- A **transversal** could be any line that intersects two or more lines.

Check out this diagram showing two lines (*l* and *m*) cut by a transversal (*n*). Also note that while Lines *l* and *m* look parallel in our picture they don't always have to be so.

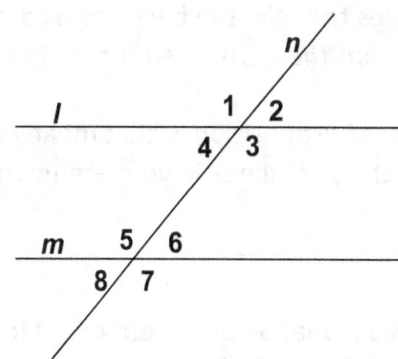

- Angles 1 and 5 are **corresponding angles**. Angles 2 and 6 are also corresponding angles. *Which other pairs of angles are corresponding angles?*

- Angles 4 and 6 are **alternate interior angles**. So are Angles 3 and 5.

- Angles 1 and 7 are **alternate exterior angles**. So are Angles 2 and 8.

Which pairs of angles on the above picture are vertical angles?

Euclid proved many theorems about lines and angles. Let's have a look at the form of such an argument now. The idea for the proof is slick – and it had to be: Euclid didn't have much machinery built up to use.

Theorem (Postulate 5): In an isosceles triangle, the base angles are congruent.

Suppose that △ABC is an isosceles triangle.

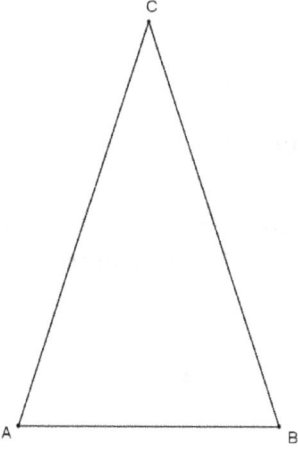

[Notice that we began by stating what is assumed and we drew a picture with labels to help others follow along. This is a good practice that you should do also.]

Now, we know segment *AC is congruent to BC*.

[Because the triangle is isosceles.]

We also know that segment *CB is congruent to CA*.

[Strange. We know. Just bear with us.]

Also, ∠ACB is congruent to ∠BCA (because they are the *same* angle,) and AB is congruent to BA. So, △ABC is congruent to △BAC by Proposition 4. Therefore, ∠CAB is congruent to ∠CBA, and we are done.

[Notice how we set it up to compare the triangle to itself- but backwards – so we could use Proposition 4. This was Euclid's slick idea.]

Don't worry. We aren't often this clever and we don't expect you to be either. But we *will* ask that you try to make some of the more straightforward arguments.

Homework:

You always pass failure on the way to success.

Mickey Rooney

1) Go back and do all the things in italics in the *Read and Study* and the *Connections* sections.

2) The sixth Standard for Mathematical Practice from the Common Core State Standards argues in part that "mathematically proficient students try to communicate precisely to others. They try to use clear definitions in discussion with others and in their own reasoning … By the time they reach high school they have learned to examine claims and make explicit use of definitions." In other words, knowing and understanding precise definitions is a very important mathematical practice. Make yourself a definitions quiz and learn the bolded and underlined terms in this section.

3) Read the following selection of *Euclid's Propositions* from the appendix and draw a notated sketch for each to help you understand what it is saying. Identify what is given (assumed) in the statement and what is concluded by the statement.

    a. Proposition 13
    b. Proposition 14
    c. Proposition 15
    d. Proposition 27
    e. Proposition 28
    f. Proposition 29
    g. Proposition 30

4) Prove Proposition 15: If two straight lines cut one another, then they make vertical angles equal to one another.

5) Prove Proposition 30, that straight lines parallel to the same straight line are parallel to each other.

Class Activity 5: Enough is Enough

> *I learned very early the difference between knowing the name of something and knowing something.*
>
> Richard Feynman

Suppose you are given some information about a triangle *ABC*. In which of the following cases will the information be enough to allow you to determine the exact size and shape of the triangle? If you have enough information, draw a triangle guaranteed to be exactly the same size and shape as $\triangle ABC$. If you do not have enough information, describe the problem you encounter in attempting to draw $\triangle ABC$.

You will need to use a ruler to measure lengths in centimeters (cm) and a protractor to measure the angles in degrees.

a. $\overline{AB} = 4$ cm and $\overline{BC} = 5$ cm

b. $\overline{AB} = 8$ cm, $\overline{AC} = 6$ cm, and $\angle BAC = 45°$

c. $\overline{AB} = 8$ cm, $\overline{AC} = 7$ cm, and $\angle ABC = 45°$

d. $\angle ABC = 75°$, $\angle BCA = 80°$, and $\angle CAB = 25°$

e. $\overline{BC} = 7$ cm, $\overline{AC} = 8$ cm, and $\overline{AB} = 9$ cm

f. $\overline{AB} = 9$ cm, $\overline{BC} = 3$ cm, and $\overline{AC} = 4$ cm

g. $\overline{AB} = 7$ cm, $\angle ABC = 25°$, and $\angle BAC = 105°$

h. $\overline{BC} = 11$ cm, $\angle ABC = 75°$, and $\angle BAC = 40°$

Read and Study

Pure mathematics is, in its way, the poetry of logical ideas.
Albert Einstein

We have been talking about triangles without officially defining them, or any other polygon, for that matter. Let's fix that now.

A **polygon** is a simple, closed curve in the plane made up entirely of line segments. The line segments are called sides and the points where segments meet are the vertices. Let's have a closer look at the pieces of this definition. First of all a mathematician uses the word "curve" to talk about pretty much any pencil line you could draw without lifting your pencil from a paper. A curve need not be curvy; it could even be perfectly straight.

A curve in the plane is **simple** if it has no loops and no branches. A curve is **closed** if it has a boundary that separates outside from inside. *Decide whether each of the following is a polygon based on the definition* (this is how we always make such decisions in mathematics).

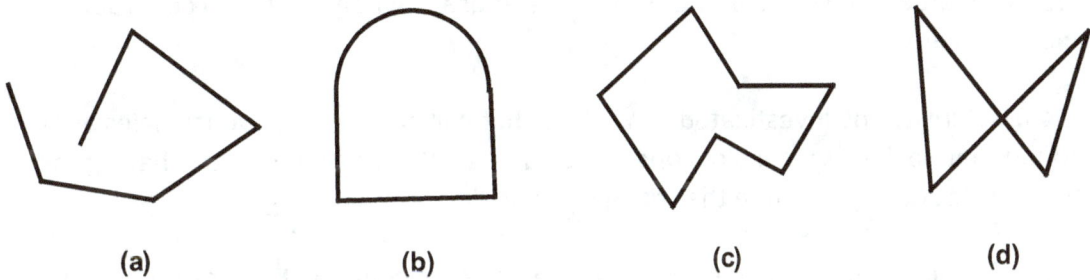

(a)   (b)   (c)   (d)

Do you see that only c) is a polygon? a) is not closed. b) is not made only of line segments. d) is not simple. Notice too that a solid object like this triangle below is not a polygon. It is the *boundary* of the object that is a polygon.

A **triangle** is a polygon with exactly three sides. A triangle is **equilateral** if all its sides are congruent. If only two sides are congruent, then we say it is **isosceles**. If no sides are congruent, then it is **scalene**. A triangle with an angle bigger than a right angle is called **obtuse**. If all its angles are less than a right angle, we say the triangle is **acute**.

Now it is time to talk about the idea of *congruence*. You were working with this idea in the *Class Activity*.

For now we will use the following definition: two geometric objects are **congruent** if they can be moved so that they coincide (sit on top of one another and fit exactly). We will make this definition more precise later in the book.

The idea of congruence is related to the idea of equality, but it is not the same thing. Congruence is a relationship between *objects* whereas equality is a relationship between *numbers*.

We would say two line segments are *congruent* (coincide), and we would say that the measures of their lengths (numbers) are *equal*.

We would say that two angles are *congruent* (coincide), and we would say their measures in degrees (numbers) are *equal*.

We do not use the equals sign (=) for congruence. Instead, we have a special symbol ($\cong$) to say that $\overline{AB}$ is congruent to $\overline{CD}$, ($\overline{AB} \cong \overline{CD}$). Be sure to use $\cong$ when you mean congruence and = when you mean equality.

In the *Class Activity* you investigated conditions that will ensure that two triangles are congruent. You found that having two pairs of congruent sides is not sufficient, but that having three pairs of congruent sides does guarantee the triangles coincide.

The case of angles is more complex. Having three pairs of congruent angles is *not* sufficient information, but if we have two pairs of congruent angles and one pair of congruent sides, we do get congruent triangles. And then there is the case where we have two pairs of congruent sides and one pair of congruent angles – sometimes we have congruent triangles and sometimes not – it makes a difference whether or not the angles in question are the angles between the two pairs of congruent sides.

Euclid compiled all of this information about when triangles are congruent into these four theorems. Notice that each of these theorems is in the "if-then" statement form. *Read each one carefully – make certain you understand all the terms and can explain each one in your own words. Then sketch a picture to illustrate what each one is saying.*

1. **Angle-Side-Angle Triangle Congruence (ASA):** If two angles and the included side of one triangle are congruent to two angles and the included side of another triangle, then the triangles are congruent.

2. **Side-Angle-Side Triangle Congruence (SAS):** If two sides and the included angle of one triangle are congruent to two sides and the included angle of another triangle, then the triangles are congruent.

3. **Side-Side-Side Triangle Congruence (SSS):** If three sides of one triangle are congruent to three sides of another triangle, then the triangles are congruent.

4. **Angle-Angle-Side Triangle Congruence (AAS):** If two angles and the side opposite one of them in one triangle are congruent to the corresponding parts of another triangle, then the triangles are congruent.

You might have noticed that Theorem 4) is redundant to Theorem 1), in light of a fact we've already established about triangles in a previous section. *To what fact are we referring?*

We want you to take time to relate these theorems to the cases you investigated in the Triangle Exploration Activity. In fact, we are going to give you space here to revisit each set of conditions and decide which, if any, of the above four theorems apply to the given cases. Really do this.

a. $\overline{AB} = 4$ cm and $\overline{BC} = 5$ cm

b. $\overline{AB} = 8$ cm, $\overline{AC} = 6$ cm, and $\angle BAC = 45°$

c. $\overline{AB} = 8$ cm, $\overline{AC} = 7$ cm, and $\angle ABC = 45°$

d. $\angle ABC = 75°, \angle BCA = 80°,$ and $\angle CAB = 25°$

e. $\overline{BC} = 7$ cm, $\overline{AC} = 8$ cm, and $\overline{AB} = 9$ cm

f. $\overline{AB} = 9$ cm, $\overline{BC} = 3$ cm, and $\overline{AC} = 4$ cm

g. $\overline{AB} = 7$ cm, $\angle ABC = 25°,$ and $\angle BAC = 105°$

h. $\overline{BC} = 11$ cm, $\angle ABC = 75°,$ and $\angle BAC = 40°$

You may have noticed that there is no Angle-Side-Side congruence theorem. This is because having a congruent angle and two congruent sides (unless the angle is between the two sides) is *not* enough to guarantee congruence. Here's the problem. Consider two triangles that each has a side that measures 4 cm, another side that measures 1.5 cm, and an angle that measures 15 degrees.

Here are two *different* triangles that meet those conditions:

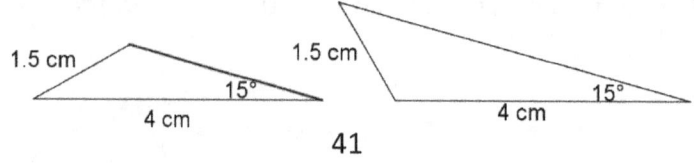

In other words, there are sometimes two choices for the third side length.

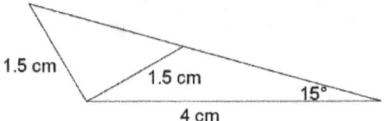

Notice too that there is no Angle-Angle-Angle congruence theorem. *Explain why not.*

Okay, now that you know the triangle congruence theorems, let's take a look at some of the other nice theorems about triangles:

5. The sum of the angle measures in any triangle is 180 degrees. (You already proved this.)

6. In a triangle, angles opposite congruent sides are congruent.

7. In a triangle, sides opposite congruent angles are congruent.

Theorems 6) and 7) are often called the **Isosceles Triangle Theorems** – and they are quite useful. *Draw a sketch of each to be certain you understand what they say.*

Notice that 6) and 7) are *not* theorems about two *different* triangles being congruent. Both theorems talk about a single triangle in which either two sides of that triangle are congruent or two angles of that triangle are congruent.

Here is one more idea that is commonly used as part of triangle congruence proofs:

8. **Corresponding parts of congruent triangles are congruent.**

We will use theorem 8) almost every time we make an argument involving triangles from now on. Since, by our definition of congruent objects, two congruent triangles coincide, every pair of corresponding parts or measurements of an **attribute** must be identical. So, in two congruent triangles, the smallest angles (if there are smallest angles) are congruent, and the longest sides (if there are longest sides) are congruent, and so on. You may (fondly) recall the anagram for this theorem, CPCTC, from a high school geometry course.

Connections to the Elementary Grades:

Who so neglects learning in his youth,
loses the past and is dead for the future.

Euripides

The Common Core State Standards asks that students in grade seven explore when given information is enough to specify a triangle. In other words, they advocate that those students should have an intuitive introduction to the triangle congruence theorems we have discussed above.

> 7.G.2 Draw (freehand, with ruler and protractor, and with technology) geometric shapes with given conditions. Focus on constructing triangles from three measures of angles or sides, noticing when the conditions determine a unique triangle, more than one triangle, or no triangle.

Copyright 2010. National Governors Association Center for Best Practices and Council of Chief State School Officers. All rights reserved.

Homework

Home computers are being called upon to perform many new functions, including the consumption of homework formerly eaten by the dog.

Doug Larson

1. Do all the italicized things in the *Read and Study* section.

2. Study each bold and underlined term used in this section. This means you should be able to explain the definition using good mathematical language and that you should be able make examples and non-examples of each term.

3. For each of Theorems 1), 2), 3), 5), 6), and 7) of *Read and Study*, find Euclid's corresponding proposition in the Appendix. (Euclid did not state Theorems 4) or 8).)

4. At which step do you know enough to draw a triangle that is congruent to the one we are describing? Explain your answer.
    I. One of the sides is 8 cm long.
    II. One of the sides is 4 cm long.
    III. The angle between the sides mentioned above is 60 degrees.
    IV. The triangle has a 90 degree angle.

5. At which step do you know enough to draw a triangle that is congruent to the one we are describing? Explain your answer.
    I. One of the sides is 3 cm long and another is 7 cm long.
    II. The angle between the 7 cm side and the unknown side is 20 degrees.
    III. The unknown side is the longest side.
    IV. The triangle has an obtuse angle.

6. At which step do you know enough to draw a triangle that is congruent to the one we are describing? Explain your answer.
    I. One of the angles measures 140 degrees.
    II. Another of my angles measures 25 degrees.
    III. One of my sides measures 7 cm.
    IV. My longest side measures 7 cm.

7. Make a mathematical argument that a triangle can only have one obtuse angle.

8. Make a mathematical argument that the two acute angles of a right triangle are complementary.

9. Make a mathematical argument for Theorem 7), that in a single triangle, the angles that are opposite the congruent sides must be congruent.

10. Make a mathematical argument that two acute angles of an isosceles right triangle are each 45°.

11. Make a mathematical argument that each angle in an equilateral triangle is 60°

12. According to the Common Core State Standards, students in grade eight should be able to do the following. Read this standard – then do the activity described in their example.

> 8.G.5 Use informal arguments to establish facts about the angle sum and exterior angle of triangles, about the angles created when parallel lines are cut by a transversal, and the angle-angle criterion for similarity of triangles. *For example, arrange three copies of the same triangle so that the sum of the three angles appears to form a line, and give an argument in terms of transversals why this is so.*

Copyright 2010. National Governors Association Center for Best Practices and Council of Chief State School Officers. All rights reserved.

13. Prove Proposition 32, namely, that the sum of the three interior angles in a triangle is two right angles. Recall that you can use any of the propositions that come before 32. We suggest that you first draw any triangle and then construct a line parallel to one of the sides of the triangle, through the opposite vertex.

14. A polygon is **convex** if all of its diagonals lie in the interior of the polygon. A **diagonal** of a polygon is a line segment that joins two non-adjacent vertices. A polygon is **concave** if it is not convex. *Use these definitions* to decide whether each polygon below is convex or concave. In each case, explain your thinking.

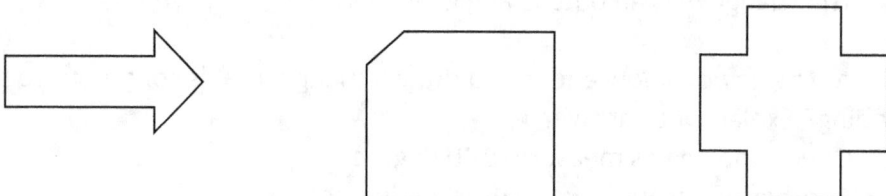

Class Activity 6: Construction Zone

The human mind has first to construct forms, independently, before we can find them in things.

Albert Einstein

1) Show that it is possible to **construct** a **ray** which **bisects** angle *ABC*. What proposition is this? Check Appendix A to see. Then prove that your construction works.

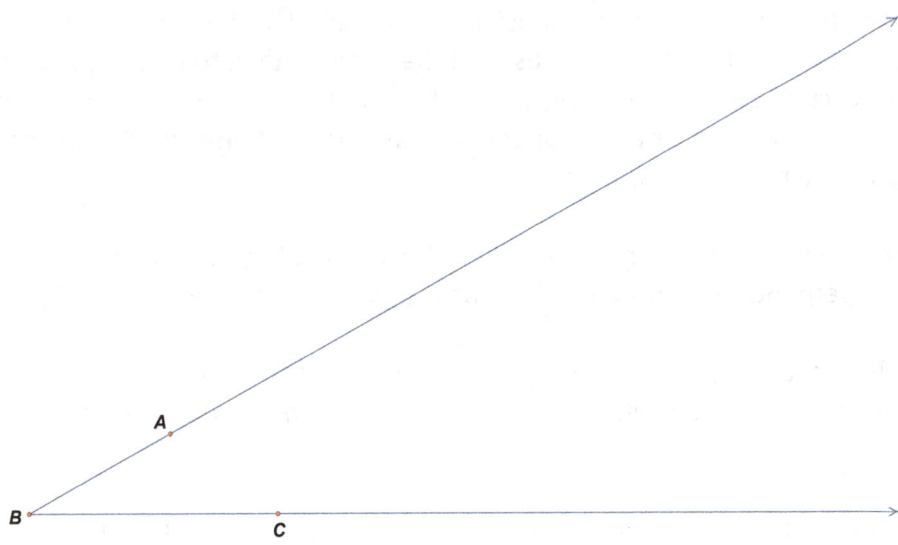

2) Show that it is possible to construct a line which both bisects and is perpendicular to line segment $\overline{AB}$. (We call such a line the **perpendicular bisector** of $\overline{AB}$.) Which proposition is this? Check to see. Then prove that your construction works.

Read and Study:

The mathematician is entirely free, within the limits of his imagination, to construct what worlds he pleases.

John William Navin Sullivan

To **construct** a geometric object is to create it using only straight line segments and circles (Euclid's first, second, and third axioms). The tools we use are the straightedge, to make line segments, and the compass, to make circles (or arcs of circles). In fact, as we mentioned earlier, you can think of your straightedge as your line-maker and your compass as your circle-maker. You cannot measure anything with a ruler or a protractor as part of your construction.

To give you an example of how mathematicians describe and justify constructions, we will show it is possible to drop a perpendicular to a given line through a given point not on the line.

Suppose we have line (*n*) and a point not on the line (*P*). It is possible to construct a line through *P* that is perpendicular to line *n* (Proposition 12). *Take out your straight edge and compass and follow along.*

First we use the compass to construct a circle centered at *P* that intersects line *n* in two points we can call *A* and *B* (we just need to draw the arc containing *A* and *B*).

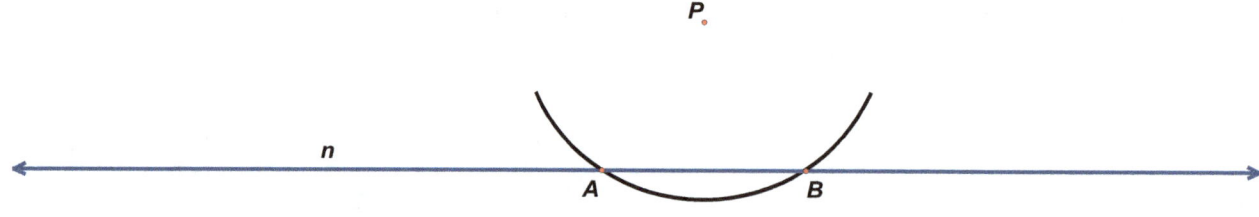

Notice that *AP* and *BP* are both radii of this circle (and thus $\overline{AP} \cong \overline{BP}$). Now we will use this same radius and construct two circles, one centered at *A* and one centered at *B*. (Again, we only need to draw the arcs of these circles that intersect below line *n*.) Label the point of intersection of these two circles *C*.

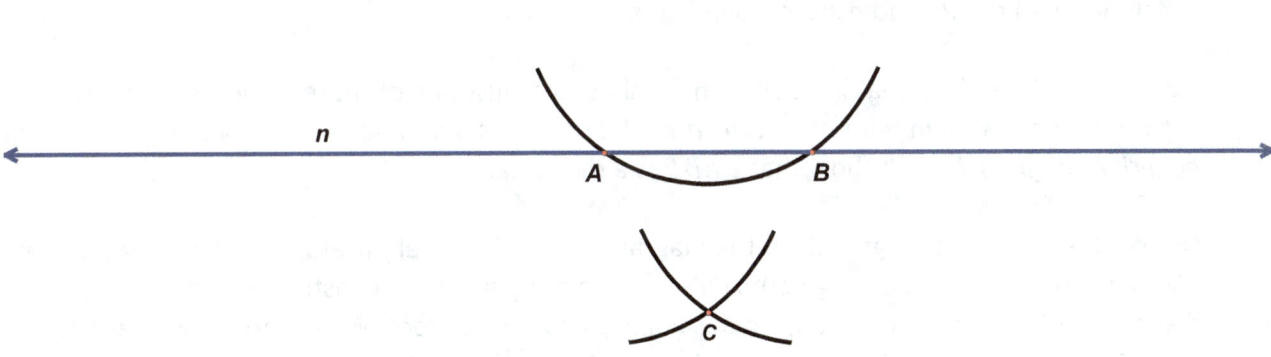

The final step in our construction is to draw the line connecting points *P* and *C* with a straightedge. This line will be perpendicular to line *n*.

We just described the "how to" of the construction of a perpendicular line. It's important to be able to carry out this procedure as there will be many occasions on which you will need to construct perpendicular lines in this class.

It is even more important to understand *why* we claim that this procedure produces perpendicular lines. We call "explaining *why* it works" **justifying (or proving) the construction**. Recall that any postulate as well as any proposition numbered below Proposition 12 is fair game for our use.

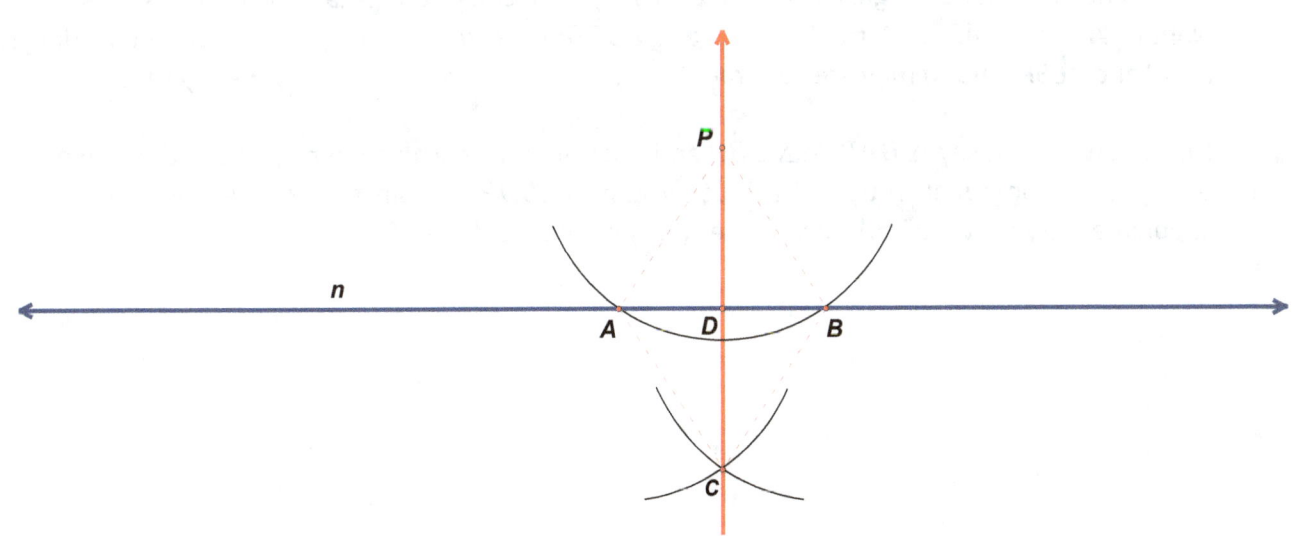

Look again at our construction diagram. We claim that the line *CP* is perpendicular to line *n*. How can we justify this claim? Well, we know that perpendicular lines are lines that intersect at right angles. So, if we can show that ∠ADP, ∠PDB, ∠BDC, and ∠CDA are each right, then we can conclude that lines *PC* and *n* are perpendicular.

Furthermore, we claim that it is sufficient to show that just one of these angles is right. (*Make an argument for this claim right now. Why are all four angles right angles if just one angle is known to be a right angle?*) We will show that ∠ADP is a right angle.

Notice the dashed line segments in the diagram. They will be very useful in making our argument. "Adding" extra lines or line segments (which were not part of the construction process) to a diagram is often a helpful strategy in designing a geometric proof. We observed before that *AP* and *BP* are both radii of the same circle (and so $\overline{AP} \cong \overline{BP}$ by the way we did the construction). Now, we can also realize that *AC* and *CB* are radii of circles congruent to the first circle. Thus, all four dashed line segments are congruent by construction and we made each of them using Postulate 3. *Make certain you can explain this part of the proof in your own words.*

These four congruent line segments form two triangles, △CAP and △CBP, that share a common side *CP*. So now we can say that these triangles are congruent by Proposition 8 (Some of you may know this as the Side-Side-Side triangle congruence theorem). You might be asking why we want to talk about △CAP and △CBP. We said we wanted to prove ∠ADP is a right angle and ∠ADP isn't even a part of △CAP or △CBP.

Well, let's take a look at △DAP and △DBP which do contain ∠ADP (and ∠BDP). If we could show these two triangles were congruent, we would be making progress toward our goal. *Why?* We have already noted that $\overline{AP} \cong \overline{BP}$. *DP* is a common side. We could use the Proposition 4 if we knew that the included angles, ∠APD and ∠BPD, were congruent. Aha, now it makes sense to want to know △ CAP ≅ △ CBP. ∠APD and ∠BPD are corresponding parts of congruent triangles △CAP and △CBP, and so they are congruent. *Make sure you understand why we say this.*

Okay, now we can say △ DAP ≅ △ DBP and, therefore, all corresponding parts of these two triangles are congruent. Thus ∠ADP is congruent to ∠BDP, and since these two angles are supplementary they are both right angles (*Why?*). We are done!

Connections to Teaching:

In grades 6-8 all students should precisely describe, classify, and understand relationships among types of two- and three-dimensional objects using their defining properties.

National Council of Teachers of Mathematics

The geometry you will teach in elementary and middle school is the geometry of Euclid. The focus however is not an axiomatic development of the subject but rather a hands-on intuitive approach. You will focus on classification and properties of 2- and 3-dimensional objects; transformations and symmetry, describing spatial relationships using maps and coordinate geometry, and geometric problem solving.

Homework:

One day of practice is like one day of clean living. It won't do you any good.

Abe Lemons

1) Do all the italicized things in the *Read and Study* section.

2) Carefully read through *Euclid's Propositions* from the appendix. Which ones are about constructing objects? What does each make?

3) Write a clear and complete description of the steps you used for each of the constructions in the *Class Activity*.

4) Justify your construction #1 in *Class Activity*, that is, prove that the ray you constructed creates two congruent angles, each half the measure of $\angle ABC$.

5) Justify your construction #2 in the *Class Activity*, that is, prove that the line you constructed is perpendicular to $\overline{AB}$ at the midpoint of $\overline{AB}$.

6) Is it possible to bisect a *line*? Why or why not?

7) You may have noticed that Euclid did not talk about measuring angles in "degrees" as we often do. We think of a full turn as being split into 360 little angles – each called a degree. We don't know when thinking in degrees began or which civilization began it – but we do have some ideas about why 360 was chosen. Why is 360 such a good choice?

8) Construct a line segment *BC* so that it is congruent to *AB* and the measure of ∠*ABC* is half of a right angle. (You don't get to use a protractor here.)

9) Prove that every point on the perpendicular bisector of a line segment is equidistant from the endpoints of that segment.

10) Prove that every point on the angle bisector of an angle is equidistant from the rays that form that angle. In other words, prove that $\overline{FD}$ is congruent to $\overline{ED}$.

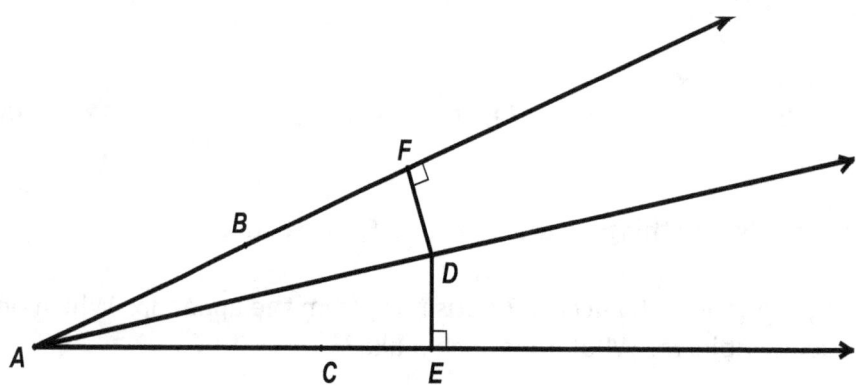

Class Activity 7: If You Build It

The shortest distance between two points is under construction.
Noelie Alito

In this activity you will perform and then justify two more constructions of Euclidean geometry. These constructions, will give you some tools with which to construct other objects later on in the course. As usual, in your justification, you can use any postulate or proposition that comes before the one you are trying to prove.

1) Show it is possible to copy a given angle so that the ray below is one of the rays of the angle. Then justify that you have done so. This is Euclid's Proposition 23.

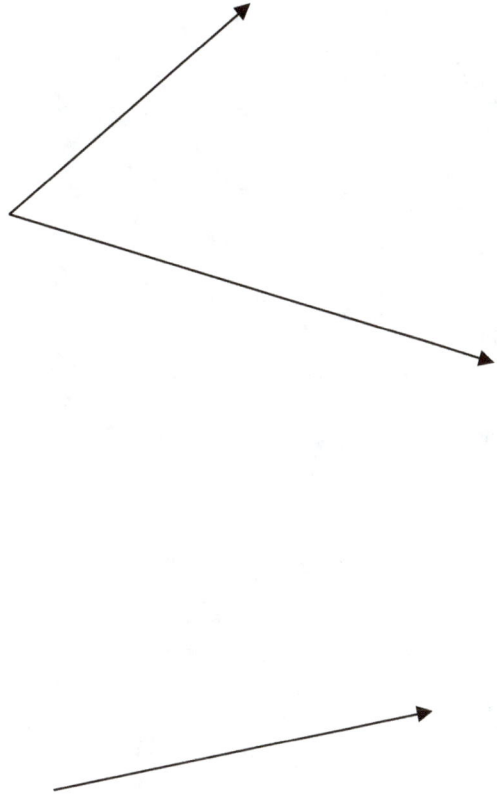

(This activity is continued on the next page.)

2) Given a line and a point on the line, show that it is possible to construct, through the point, a line that is perpendicular to the given line, and then justify that you have done so. This is Euclid's Proposition 11.

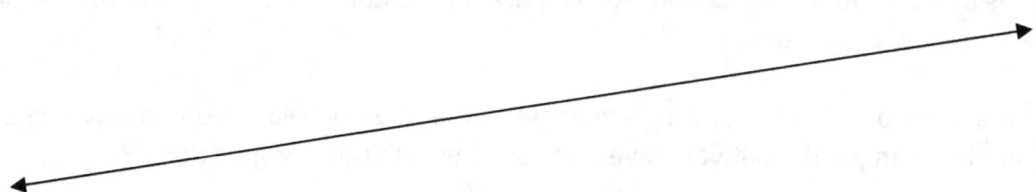

Summary of Big Ideas from Chapter One

Hey! What's the big idea?

Sylvester

- One definition of **geometry** is that it is the study of ideal shapes and the relationships that exist among them.

- A second definition is that **geometry** is an axiomatic system about objects called "points," collections of points called "lines," and the relationships between points and lines.

- Axiomatic systems define the rules governing the particular geometry.

- Proven consequences of a particular set of axioms are theorems.

- A finite geometry consists of a finite number of objects and their relationships.

- Mathematicians are very careful about distinctions within mathematical language.

- Congruence is an important relationship between geometric objects. We say two objects are congruent if they coincide when placed on top of each other.

- We construct an object by creating it using only straight line segments and circles.

- *You* will represent the mathematical community for your students. They will look to you to understand what we mathematicians do and how we think. Your students will try to discern the meanings that you, your curriculum materials, and other students give to ideas, strategies, and symbols through their participation in doing mathematics in your classroom. You must be aware that talking about the meanings of words, ideas, and symbols is an important part of your role as a teacher; and you must be transparent and careful in your use of words, symbols, and notation when you are teaching mathematics.

Chapter Two

Learning and Teaching Euclidean Geometry

Class Activity 8: Circular Reasoning

Nature is an infinite sphere of which the center is everywhere and the circumference nowhere.

Blaise Pascal

For this activity, each person in your group will need to draw three points of a blank sheet of paper as follows: One person should arrange the points so that the triangle formed with the points as its vertices is an acute scalene triangle. Another should arrange an obtuse triangle. Another should arrange a right triangle, and if there is a fourth person, that person should arrange his or her points to make an equilateral triangle. Make your triangles fairly large, you are going to be doing lots of constructing. First, a few definitions:

The **circumcenter** (C) of a triangle is intersection point of the perpendicular bisectors of the sides.

The **incenter** (I) of a triangle is the intersection point of the angle bisectors.

The **orthocenter** (O) is the intersection of the altitudes (heights) of the triangle.

The **centroid** (M) is the point of intersection of the medians (lines joining a vertex with the midpoint of the opposite side) of the triangle.

Carefully construct each of these centers for your triangle. (You may want to use different colored pencils for different constructions.) Then label each special point. When you are done, compare your results and answer the following questions:

1) One of these points is special because it is the center of mass of the triangle (the balancing point). Which one and why?

2) One of these points is special because it is the center of the circle containing all the vertices of the triangle. Which one and why?

3) One of these points is special because it is the center of the biggest circle that can be placed inside the triangle. (The circle that is tangent to all three sides.) Which one and why?

4) Which three of the four special points always lie on the same line?

5) Which of the points could lie outside of the triangle? For what type of triangles does that happen? Why does this make sense?

Read and Study:

It is easier to square the circle than to get round a mathematician.
Augustus De Morgan

We now have some basic tools with which to study Euclidean Geometry, and in this chapter we will develop even more.

A mathematical **circle** is the set of points that are equidistant from a given point, called the **center** (*O* in the diagram below). The diagram shows some of the other important terms associated with a circle. *Be certain you understand each term and can explain its mathematical definition.*

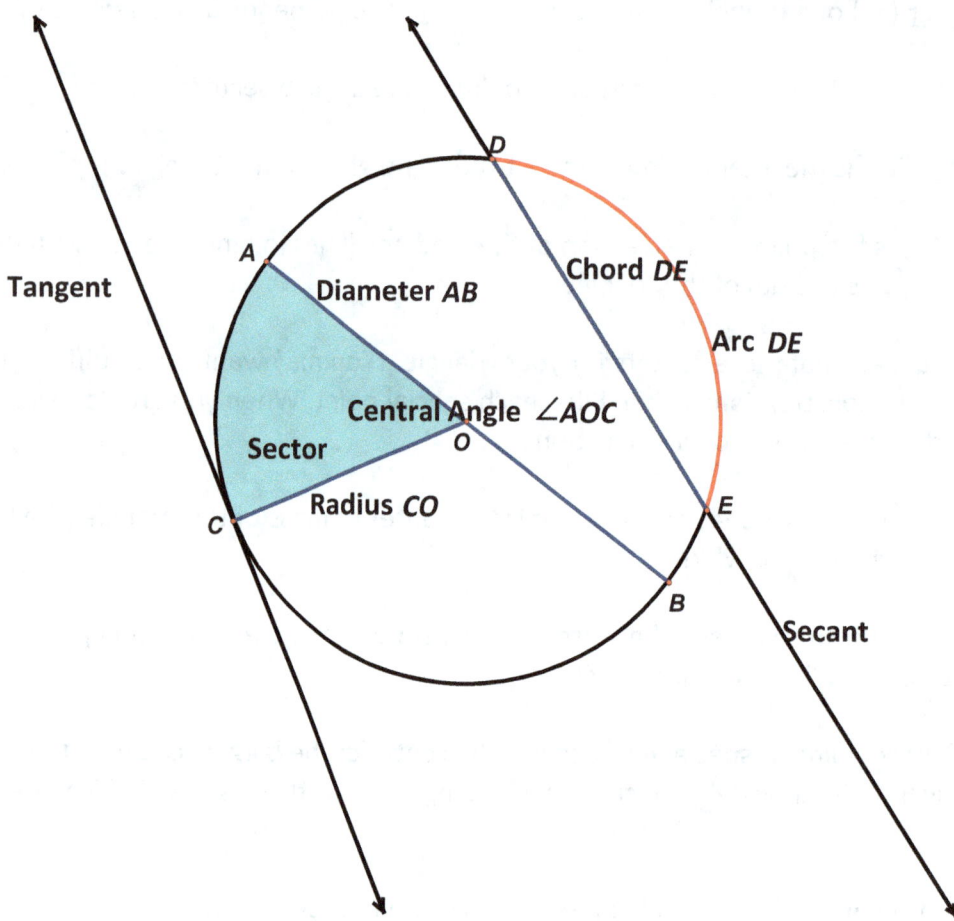

It is an amazing fact that *for any size circle*, the ratio of the circumference to the diameter is constant. We now call this constant **pi (p)**. Over the years many mathematicians have tried to find approximations for pi. Archimedes, a genius of the Greek mathematicians, found approximate bounds for its value using circumscribed and inscribed polygons with 96 sides (he proved that $\frac{223}{71} < \pi < \frac{22}{7}$. (The average of these two values is roughly 3.1419, a pretty darn good estimate.) It is worth noting that even when we use the p key on a calculator or ask a computer to compute it, we are using an *approximate* value because p is not rational and therefore does not have a decimal name that terminates or repeats. Students commonly use either 3.14 or $\frac{22}{7}$ as an approximate value for p when carrying out calculations involving circles.

Euclid's work contains many theorems about circles. We will discuss two of them now and ask you to explore some more in the *Homework* section. The first theorem we'll look at says that two chords of a circle are congruent if and only if their corresponding arcs have the same measure.

First, that "if and only if" phrase means that we are getting two theorems for the price of one. Both the statement and its converse are true. Thus, this theorem gives us two if-then statements: if two chords of a circle are congruent then their corresponding arcs have the same measure, and if two arcs of a circle have the same measure, then their corresponding chords are congruent.

Let's illustrate these theorems in a diagram: if chords $\overline{AB}$ and $\overline{CD}$ are congruent, then the arcs AB and CD (shown in dark red) are also congruent, and vice versa.

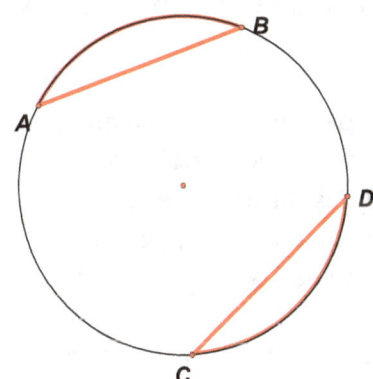

Our second theorem is one about inscribed angles which states that the measure of an angle inscribed in an arc is one-half the measure of its intercepted arc. To make sure we understand what this theorem is saying, we need distinguish between an **inscribed angle**, an **intercepted arc**, and a central angle. In the following diagram, ∠ADB is inscribed in arc ACB which is called its intercepted arc. Arc ACB is measured by the central angle ∠AOB. *Restate the theorem in terms of ∠ADB and ∠AOB.*

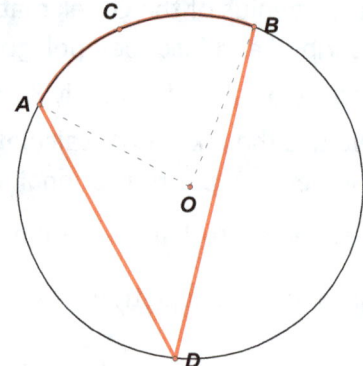

Note that we have not proved either theorem in this section, but take a few minutes now to do some measurements so that you can see that they might be true.

Now take out your compass and straightedge and follow along. Write down two distinct points and name them *A* and *B*. Construct *Circle AB* with center at *A* and point on the circle *B*. Construct any chord (and name it *PQ*) on circle *AB*.

Construct a line through *A* that is perpendicular to chord *PQ* and label the point of intersection of *PQ* and this perpendicular line *M*. Think about what would happen if you could "move" *P* and *Q* around on the circle? In other words what happens as you change the positions of *P* or *Q* and keep point *M* as the intersection point of that new *PQ* and the line that is perpendicular to *PQ* through *A*? What do you notice about *M*? Why do you think this happens? Can you make a general argument to support your conjecture?

Find the intersection points of the perpendicular line and circle *AB* and call them *R* and *S*. Construct line segment *RS* and erase the perpendicular line and point *M*. This kind of a segment with endpoints on the circle that goes through the center of the circle is called a **diameter**. *For any chord PQ that is not a diameter, what can we say about the chord's length in comparison to the length of any diameter for a given circle?*

Construct the perpendicular bisector of chord *PQ*. Imagine moving points *P* and *Q* around on the circle. *What happens to the resulting perpendicular bisector?*

Now, construct a new *Circle AB* (again with center *A* and point on circle *B*). Construct a diameter of the circle *AB* with endpoints *B* and *C*. Pick and label a point *E* on the circle. Make segments *EB* and *EC*. "Move" *E* around the circle. *What can you conjecture about triangle BEC?*

You are going to need a final new Circle *AB*. Pick a point on the circle and label it *P*. Construct radius *AP*. Construct a line through *P* perpendicular to segment *AP*. Imagine moving *P* around the circle. *Is this perpendicular line a* **secant** *or a* **tangent**?

In book four of *Elements*, Euclid proved several theorems about circles, one of which is that three distinct, non-collinear points determine a unique circle (one that passes through all three points).

You constructed that circle when you found the circumcenter of your triangle. In the *Homework* section, you will justify that construction.

Connections to Teaching:

Geometry is a natural place for the development of students' reasoning and justification skills.

NCTM, Principles and Standards, 2000

Perhaps the best regarded model regarding children's geometric reasoning is the **van Hiele Levels**. Pierre van Hiele and Dina van Hiele-Geldof were husband and wife, educators, and researchers of children's thinking. They asserted there are five developmental levels of geometric reasoning. Before we discuss the levels, we'll give you some general information about them according to the research. First, the levels appear to be sequential; that is, children must pass through them in order. Second, they are not so much age-dependent as experience-dependent. It is *geometric activity* at their *current* level that prepares children for more sophisticated reasoning. Finally, it appears that instruction and language at levels higher than that of the child will actually *inhibit* learning. That's a little worrisome for teachers – because it means that you can do harm if you do not tailor instruction to the specific levels of your students.

Here is the model as it is described by Battista (2007):

The van Hiele Levels of Geometric Reasoning

Level 0: **Visual**. Children recognize geometric objects by their overall appearance based on a few prototypical examples of the objects. For example, a child at this level might reject a triangle that is oriented differently than those she is used to seeing or one that is extremely long and thin. If you ask kindergartener *why* a shape is a triangle, she will likely tell you, "because it *looks* like one."

Level 1: **Descriptive or Analytic**. Children begin to identify properties of geometric objects and use appropriate terms to describe those properties. For example, a child at this level could classify triangles based on the property that triangles have exactly three sides.

Level 2: **Abstract or Relational**: Children recognize relationships between and among properties of geometric objects, and will make and follow arguments and classify shapes based on these properties.

Level 3: **Deduction**: Students construct arguments about geometric objects using definitions, axioms, and deductive reasoning. Your high school geometry course was probably taught at this level.

Level 4: **Rigor**: Students at this level will understand that there are many geometries, each with its own axiomatic system and models. In this course we will give you a sense of this.

Upper elementary and middle grades students typically test at van Hiele Level 1 or Level 2. As a teacher of these grades your job is to give your students lots experiences like the following:

- Classifying objects based on definitions. For example, you might define a **rhombus** as a **parallelogram** with four congruent sides, and ask your class to decide whether several shapes are rhombi based on that definition.

- Making and testing conjectures about geometric objects.

- Using informal deductive language, words like "all," "some," "there exists," and "if-then" statements. For example, you might ask your students to decide if the following statement is true: if a shape is a **rectangle**, then it is a rhombus. (*Is it true?*) Or you might ask a question like, does there exist a rectangle that is a rhombus? (*Does there?*)

- Exploring the truth of a statement, its converse, and its contrapositive. *For example, decide whether each of the following is true or false. Make an argument in each case.*

    If a **quadrilateral** is a square, then it has four congruent sides.

    If a quadrilateral has four congruent sides, then it is a square.

    If a quadrilateral does not have four congruent sides, then it is not a square.

- Problem solving involving geometric objects and relationships.

- Making informal deductive arguments about objects and relationships among objects.

- Making models and pictures of geometric objects.

Here is a middle grades activity focused on circles that allows students to study models, make and test conjectures, and make informal arguments. The idea is to estimate the value of π by measuring a variety of circles to find the number of times the diameter of each circle fits into its circumference. *Take a moment to do that now with the two circles below then answer the following questions:*

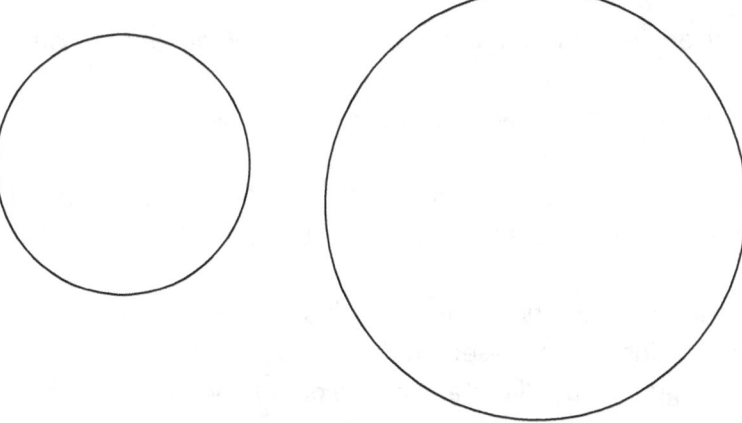

1) *Why aren't your answers exactly the same?*

2) *Does the value of π depend on the units of measurement that you use? Explain.*

What van Hiele level do children need to reach in order to do various parts of the above circle activity?

Many middle school students have heard that π is an irrational number, but they are not clear about what that *means*. It does *not* mean that the number of times the diameter of a circle fits into its circumference is changing in some way. It does *not* mean that the number of times the diameter of a circle fits into its circumference isn't an exact value. It *is* an exact value, and we call it π. It simply means that π has a decimal name that never ends nor repeats and so any way we write π with a decimal or a fraction name is merely an approximation of the number of times the diameter of a circle fits into its circumference.

Homework:

> *The knowledge of which geometry aims is the knowledge of the eternal.*
> *Plato, Republic, VII, 52.*

1) Go back and do all the things in italics in the *Read and Study* section.

2) Do all the italicized things in the *Connections* section.

3) Make yourself a definitions quiz and learn all the bolded and underlined terms in the section (including those that appear in the *Class Activity*).

4) Here is a list of activities. Classify each according to the van Hiele Level that it best fits:
   a) Sorting shapes based on the number of sides.
   b) Arguing that all rectangles are parallelograms.
   c) Identifying circle shapes in the classroom.
   d) Doing the activity on estimating π from the *Connections* section.
   e) Doing the *Two Finite Geometries Class Activity*.

5) Suppose the Earth is an ideal sphere and you have wrapped a rope tightly around the equator. Now suppose you added enough slack to raise the rope uniformly one foot off the ground all the way around the equator. How much longer rope would you need? Explain why this makes sense.

6) Prove that the circumcenter of a triangle is equidistant from the three vertices of the triangle. You will have to rely on the way you *constructed* the circumcenter. You may use any of the propositions in Book I for this argument.

7) Prove that the incenter of a triangle is equidistant from the three sides of the triangle. Again you will need to rely on how you constructed the incenter, and you may use any of the propositions in Book I for this argument.

8) Calculate the number of times the diameter of the below circle fits into the perimeter of the inscribed square and then into the perimeter of the inscribed hexagon. What is it you are doing here? If you did the same thing using a 72-sided polygon, what approximately would your answer be?

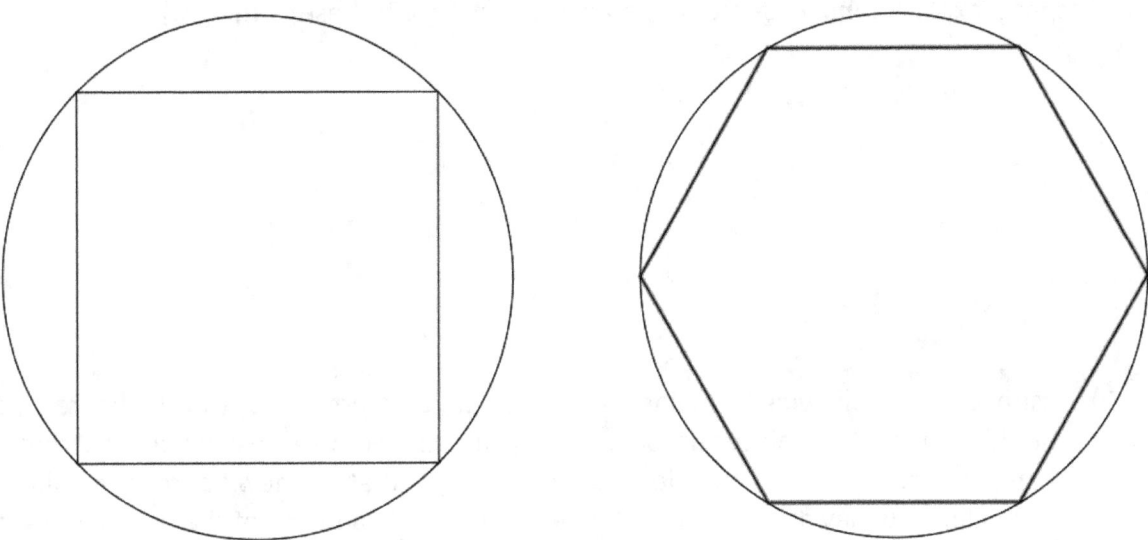

Class Activity 9: Finding Formulas

Everything should be made as simple as possible, but not one bit simpler.
                                                                Albert Einstein

1) Using the definition of **area** as the number of square units it takes to fill a two-dimensional space, explain it *makes sense* that area of a rectangle is (base) × (height).

2) Justify that the following formulas make sense. If you rearrange any of the figures, you should argue that the pieces fit together as you claim. For example, if you do something to change a triangle into a rectangle, you need to argue that the new figure is actually a rectangle. (You may assume the rectangle area formula and any of the postulates and propositions in Book I of *Elements*.)

   a) Area of a triangle = ½ (base) × (height)

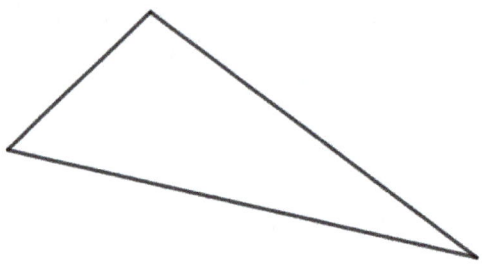

(This activity is continued on the next page.)

b) Area of a parallelogram = (base) × (height)

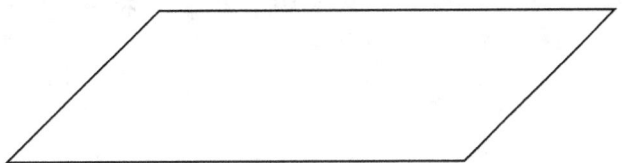

c) Area of a trapezoid = ½ (base I + base II) × (height)

Read and Study:

The description of right lines and circles, upon which geometry is founded, belongs to mechanics. Geometry does not teach us to draw these lines, but requires them to be drawn.

Issac Newton, Principia Mathematica

In *Elements* Euclid did not explicitly define length, area or volume – but it seems as though he thought of these constructs much as we do today – for example, he likely thought of "area" as the amount of two-dimensional space occupied by an object. All the definitions we will use in this section depend on comparing an object to a **unit** of measurement. In fact, they are all about how many units "fit" inside an object.

The **length** of an object is the number of 1-dimensional units (like a line segment) that fit in a 1-dimensional object.

A length unit might look like this: ____

The **area** of an object is a measure of the number of 2-dimensional units (like solid (filled-in) squares) that fit in a 2-dimensional object.

An area unit might look like this: ■

The **volume** of an object is a measure of the number of 3-dimensional units (solid cubes perhaps) fit in a 3-dimensional space.

A volume unit might look like this solid block: ⬛

This may sound simple, but we can't begin to tell you how often students are confused about this. Ask people what *area* is, for example, and most will respond that area is "base times height." But this isn't the *idea* of area, it is simply a formula for finding an area of some very specific objects (namely parallelograms: the formula *doesn't even work* for other things). When you are asked to compute an area, please don't resort to a couple of memorized formulas without thinking about what area means and whether those formulas apply, and please help your students to understand the *idea* of measurement.

Connections to Teaching:

In grades 6-8 all students should develop and use formulas to determine the circumference of circles, and the areas of triangles, parallelograms, trapezoids, and circles, and develop strategies to find the area of more complex shapes.
National Council of Teachers of Mathematics
Principles and Standards for School Mathematics, p. 240

To help your students to think of *area* as the number of solid squares that fill or cover a 2-dimensional object, you might start by having them trace the object on square-grid-paper and then ask them to estimate and then count the number of squares it takes to fill (cover) the object. Similarly, they can learn to think of volume as the number of solid cubes that it takes to fill a three-dimensional object.

Here are two of the relevant Common Core State Standards for children in grade six. *Read these carefully.*

> **6.G. Solve real-world and mathematical problems involving area, surface area, and volume.**
> 1. Find the area of right triangles, other triangles, special quadrilaterals, and polygons by composing into rectangles or decomposing into triangles and other shapes; apply these techniques in the context of solving real-world and mathematical problems.
>
> 2. Find the volume of a right rectangular prism with fractional edge lengths by packing it with unit cubes of the appropriate unit fraction edge lengths, and show that the volume is the same as would be found by multiplying the edge lengths of the prism. Apply the formulas $V = l\,w\,h$ and $V = b\,h$ to find volumes of right rectangular prisms with fractional edge lengths in the context of solving real-world and mathematical problems.

Copyright 2010. National Governors Association Center for Best Practices and Council of Chief State School Officers. All rights reserved.

In fact, a great way to help children *understand area formulas* is to have them *see* the formulas (by cutting and pasting) based on formulas they already know like you did in the *Class Activity*.

Homework:

Learning without thought is labor lost; thought without learning is perilous.
Confucius

1) Make sure that you can justify all the formulas from the *Class Activity*.

2) Find the area of the pentagon in at least 3 different ways. Each square is one centimeter long.

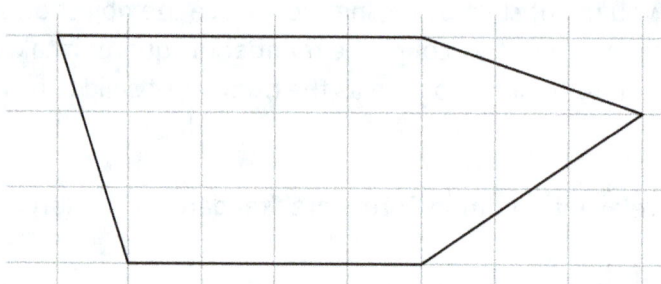

3) Middle school students should have a variety of opportunities to see why it makes sense that the area of a circle should be $\pi \times r^2$ (where *r* is the radius). Below is a picture that gives the idea of an argument for that fact. What is the idea here? Why is this just an *idea* of the argument?

Class Activity 10: Playing Pythagoras

Everything you can imagine is real.

Pablo Picasso

1) State the Pythagorean Theorem. (It's not just $a^2 + b^2 = c^2$. What are the *conditions* on *a*, *b* and *c*? You need an if-then statement.) Now, state its converse.

2) You will consider what is likely Euclid's own proof of this theorem now. We are going to explain the big ideas and your group should follow along and supply the details.

   First, Euclid claimed that ΔFBC had the same area as triangle ΔFBA (half the pink square) because both triangles have the same base and the same height. *Make sure everyone in your group sees and understands that.*

   Now Euclid argued that ΔFBC was equal to (congruent to) ΔDBA. *Make that argument.*

   Next, Euclid argued that ΔDBA had the same area as ΔBDK (half of the pink rectangle) because both have the same base and the same height. *Check it out.*

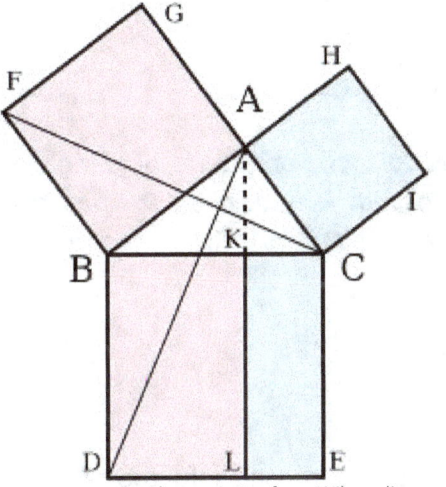

Image used with permission from Wikapedia.com

So that means that the pink square has the same area as the pink rectangle.

A similar argument shows that the blue square has the same area as the blue rectangle. *Go through the details of that now to be sure everyone understands it.* So the areas of the squares on the right triangle's sides sum to the area of the square on the hypotenuse.

Now, where in the proof did you need the fact that the triangle was a right triangle? Explain.

Read and Study:

The cowboys have a way of trussing up a steer or a pugnacious bronco which fixes the brute so that it can neither move nor think. This is the hog-tie, and it is what Euclid did to geometry.

Eric Temple Bell
In R. Crayshaw-Williams, The Search for Truth

The Pythagoreans were a group of mystics and scholars who lived in Greece about 400 BC. While there is no written record of their beliefs or work, they are thought to have ascribed to a belief in the mathematical order of the universe. They are also thought to have *proved* the theorem that bears their name – although the relationship among the sides of a right triangle was known earlier in Babylonia and perhaps in China.

The Pythagorean Theorem is a key milestone in Euclid's *Elements*. Euclid arrives at this theorem and its converse as the final propositions of Book 1. (There are 13 books that make up the *Elements*). So, we think that he must have considered it of great significance, if not the whole purpose for developing the propositions that precede it. It's a big deal because it is the key to defining Euclidean distance. We'll talk more about that later in this text.

Connections to Teaching:

I constantly meet people who are doubtful, generally without due reason, about their potential capacity [as mathematicians]. The first test is whether you got anything out of geometry. To have disliked or failed to get on with other [mathematical] subjects need mean nothing.

J.E. Littlewood, A Mathematician's Miscellany

The Pythagorean Theorem is one of those useful tools for solving problems; unfortunately, students usually remember only the $a^2 + b^2 = c^2$ part, as though it's just a formula and not a *relationship* among the areas of the squares on the sides of a right triangle. Your job is to help your students to *see* this theorem.

The initial statement of this theorem was always given in terms of *areas*. It went something like this:

Pythagorean Theorem: The square on the hypotenuse of a right triangle is equal to the sum of the squares on the other two sides.

Here is a puzzle that helps to make the points that the areas of the squares on the legs of a right triangle exactly fit to fill up the square on the hypotenuse. *Trace the diagram, then cut out the parts of the squares on the legs of the right triangle and see if you can rearrange the pieces to fit the square on the hypotenuse (Hint: the tiny square goes in the middle).*

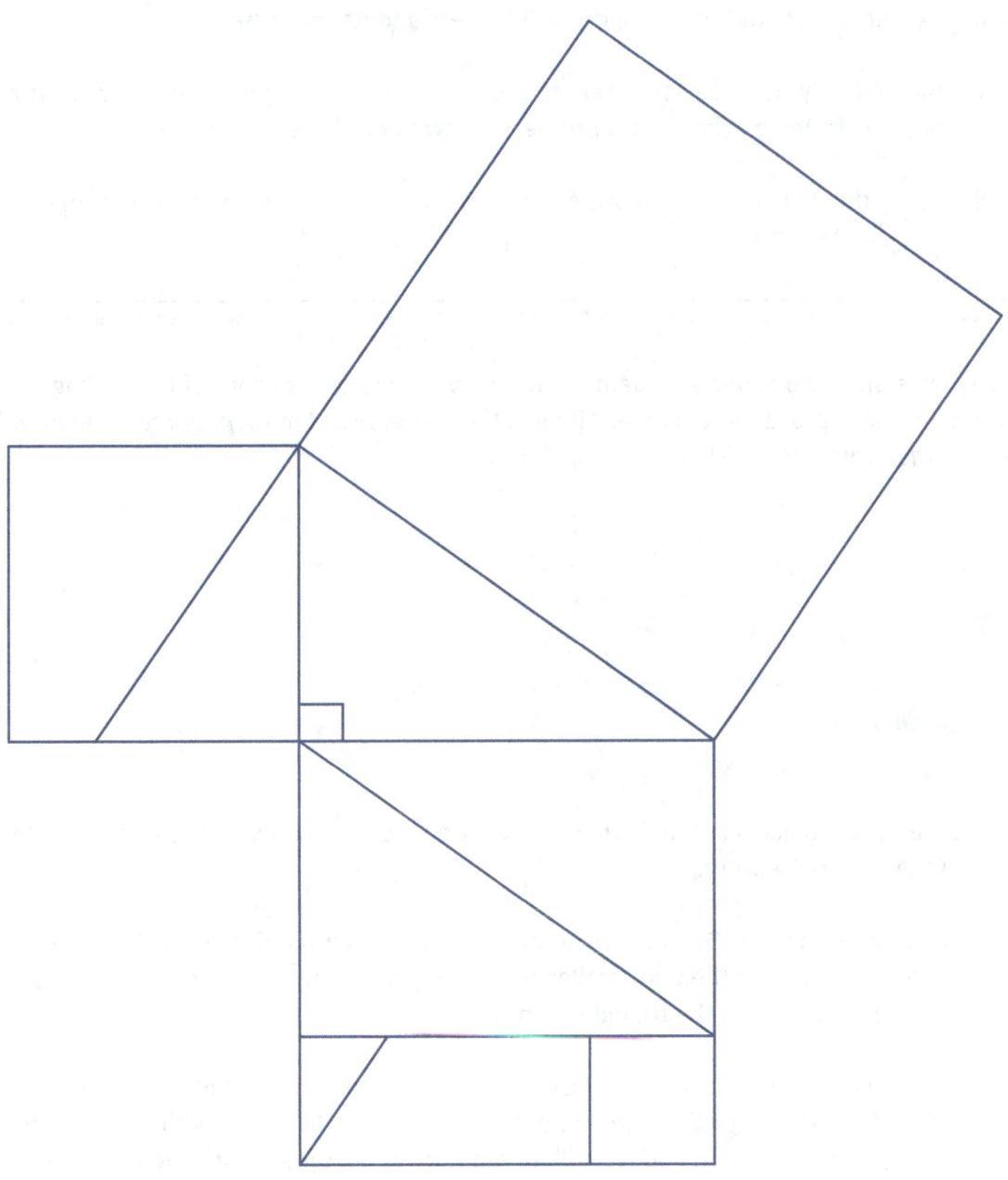

Pythagorean Puzzle

Here are three of the Common Core State Standards for children in grade eight. *Read these carefully.*

8.G. Understand and apply the Pythagorean Theorem.

6. Explain a proof of the Pythagorean Theorem and its converse.

7. Apply the Pythagorean Theorem to determine unknown side lengths in right triangles in real-world and mathematical problems in two and three dimensions.

8. Apply the Pythagorean Theorem to find the distance between two points in a coordinate system.

Copyright 2010. National Governors Association Center for Best Practices and Council of Chief State School Officers. All rights reserved.

Notice how standard 6 expects students to not only understand a proof of the Pythagorean Theorem, but also prove its converse. *What is the converse of the Pythagorean Theorem? State it carefully, then try to prove it!*

Homework:

The ability to focus attention on important things is a defining characteristic of intelligence.

Robert J. Shiller

1) Do all the problems in the *Connections* section including proving the converse of the Pythagorean Theorem.

2) The puzzle from the *Connections* section only works with a right triangle. If the triangle is acute, is the sum of the two smaller squares bigger or smaller than the square on the hypotenuse? What if the triangle is obtuse?

3) The Common Core Standards for Geometry advocate that students in eighth grade learn to do the following regarding the Pythagorean Theorem. We have addressed the first standard in this section and we will do the third in a later section when we study analytic geometry. Here are some problems to give you more practice with the second: solving real-world and mathematical problems.

   a) The school is 4 miles due east of your house and the mall is 8 miles to the north of your house. How far apart (as the crow flies) are the school and the mall?
   b) A square has a diagonal of length 10 inches. What is its area?

c) For a rectangular shoe box with sides of length *a*, *b* and *c*, explain why the diagonal *d* satisfies the "three-dimensional Pythagorean theorem" given by the equation: $a^2 + b^2 + c^2 = d^2$.

4) Study the diagram below and then use it to provide another proof of the Pythagorean Theorem. You may assume that all four triangles are congruent right triangles.

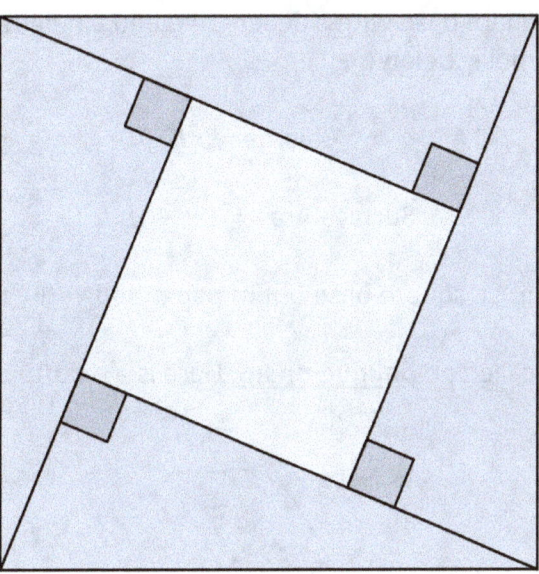

Class Activity 11: Nothing but Net

I've failed over and over again in my life and that is why I succeed.
Michael Jordan

If you have a **prism** with a square base with side length b and a height h, then its surface area and volume are given by the formulas below:

$$\text{Volume} = b^2 h$$

$$\text{Surface Area} = 2b^2 + 4bh$$

1) Build a **right prism** with a square base out of paper and verify the above formulas.

2) A non-right prism is called an **oblique prism**. Here is a picture of one:

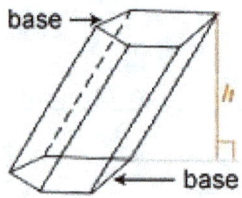

Suppose that you have an oblique prism with height h and a square base with side length b. Does the above formula for volume still hold? Build some oblique prisms and explain what you see.

Does the formula for surface area still hold? Explain.

3) See if you can make a net for an oblique (non-right cylinder) like the one shown below. What are your conjectures about the volume and surface area of an oblique cylinder compared to a right cylinder with the same height and radius?

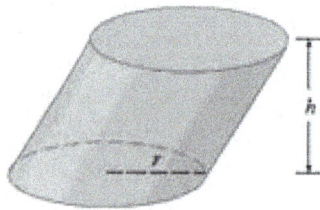

Read and Study:

You got to know when to hold 'em, know when to fold 'em ...
The Gambler by Don Schlitz

A **net** for a three-dimensional object is a two-dimensional pattern that can be folded to make the object. So, for example, here is a picture of a net that can be folded to make a cube. *Mentally fold it up*.

There are several nets that fold to make a cube. In the *Homework*, you get to find them all.

Nets are useful for studying objects like polyhedra. A **polyhedron** is a surface of a three dimensional object. In order to be a polyhedron, that surface must be closed, simple (have only one chamber), and composed entirely of polygons. The prisms and pyramids that you worked with in the class activity were both examples of polyhedra. The polygons (and their interiors) that compose the surface of the polyhedra are called **faces**. The faces meet pairwise along **edges** and the edges meet other edges at **vertices**. *How many of each: faces, edges and vertices, does the cube have?*

A **regular polyhedron** is a polyhedron made up of entirely of congruent regular polygon faces in such a way that all the vertex arrangements are the same. So the cube above is an example of a regular polyhedron because it is composed entirely of congruent square faces with exactly three faces meeting at each vertex. It turns out there are only five regular polyhedra. In order to understand this argument you will need to cut out all of the triangle, square, pentagon and hexagon faces in Appendix C and find some tape. *Take a few minutes to do those things now*.

Start with the equilateral triangles. Notice that you need to arrange at least three at a vertex in order to fold a three-dimensional object. Make the regular polyhedron that has exactly three triangle faces meeting at each vertex. It is called a tetrahedron.

Now see if you can fit four triangles at each vertex. Build that regular polyhedron. It is called an octahedron.

Finally notice that you have room to fit five triangles at a vertex and still be able to fold it up – but with six triangles at a vertex the thing lies flat on the plane and cannot be folded. So that means that there are only three regular polyhedra that can be built of equilateral triangles. Here is wire model of the regular polyhedron with five triangles at a vertex. It is called an icosahedron.

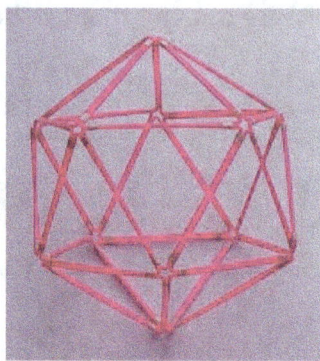

Okay. Let's move on to squares. We know we can fit three at a vertex and that gets us the cube. *Can you build something with four at a vertex? More than four? In each case, either do it, or explain why not.*

There is one more regular guy that is composed entirely of pentagons.

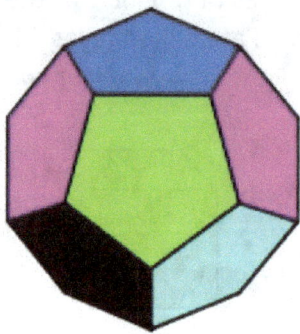

We cannot build a regular polyhedron with only hexagons because three at a vertex lie flat and cannot be folded. (*Try it.*) Polygons with even more sides than a hexagon do not work either because they cannot be folded into three dimensions. So that means there can be only five regular polyhedra. *Make sure that you understand this argument.*

Connections to Teaching:

Mathematically proficient students understand and use stated assumptions, definitions, and previously established results in constructing arguments. They make conjectures and build a logical progression of statements to explore the truth of their conjectures.
Common Core State Standards for Mathematics, p. 6.

The Common Core Standards for grade eight require that students solve real-world problems involving volume of **cylinders**, **cones** and **spheres**.

> 8.G.9 Know the formulas for the volumes of cones, cylinders, and spheres and use them to solve real-world and mathematical problems.

Copyright 2010. National Governors Association Center for Best Practices and Council of Chief State School Officers. All rights reserved.

A cylinder is similar to a prism in form. It has a circular base of radius *r* and a height *h*. You probably argued as part of the *Class Activity* that the volume of a cylinder is $\pi r^2 h$.

Now imagine a compatible cone (one with the same radius and height) living inside the cylinder. Its volume is *one third* of the cylinder's volume or $1/3\ \pi r^2 h$. This formula is difficult to derive – but you can help your students to *see* the relationship between the volumes of these objects by purchasing compatible plastic models and having the students see that it takes the water from three cones to fill the cylinder.

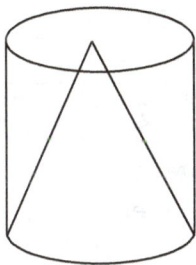

A sphere is the surface of a ball. The volume of a solid (filled) sphere is given by the formula $4/3\ \pi r^3$ where *r* is the radius of the sphere. Imagine a sphere living inside the right cylinder.

You can show your students that in order to fill the cylinder, you need the water from one compatible cone and one compatible sphere. Since the volume of the cone is $1/3\ \pi r^2 h$, the volume of the sphere must be the rest. *Do the calculation to show that the (volume of the shown cylinder) – (volume of a cone of the same height) does you give you the volume of the shown sphere.*

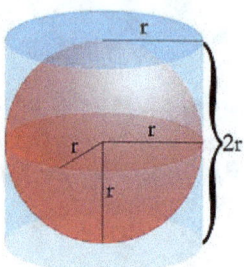

Used with permission from Wikipedia.com

You will solve some more problems involving the volumes of these objects as part of the *Homework* section.

Homework:

Doing is a quantum leap from imagining.

Barbara Sher

1) Do all the italicized things in the *Read and Study* and *Connections* sections.

2) The Common Core State Standards for students in grade six include the following:

> 6.G.4 Represent three-dimensional figures using nets made up of rectangles and triangles, and use the nets to find the surface area of these figures. Apply these techniques in the context of solving real-world and mathematical problems.

Copyright 2010. National Governors Association Center for Best Practices and Council of Chief State School Officers. All rights reserved.

Here are some problems that might meet this standard:
   a) A rectangular room is 15 feet long by 10 feet wide and has an 8 foot ceiling. Build a (scaled down) model for the room using a net.
   b) You want to paint the walls and ceiling and so need to estimate the amount of paint you will need. If a gallon of paint covers 200 square feet, how many gallons should you purchase? Explain your work.

3) Carefully make a net for a right circular cylinder. What is a formula for surface area of a right circular cylinder? What is the formula for its volume? Explain your answer in each case as you would to middle grades students.

4) If you doubled each linear dimension of your cylinder (radius and height), what would happen to the surface area? The volume? Explain.

5) How many different nets are possible for a cube that measures 1 inch on a side? Sketch them and argue that you have them all.

6) If you doubled each linear dimension of your cube (i.e., go from 1 × 1 × 1 to 2 × 2 × 2) what would happen to the surface area? The volume? Explain.

7) Are cones, spheres, or cylinders examples of polyhedra? Why or why not?

8) An ice cream snack is composed of a cone with half a sphere on top. What is the volume of the snack if the cone has radius 3 *cm* and a height of 8 *cm*?

9) Go online and search for "nets for the regular polyhedra." Print out and build each of the five. You will need these for *Class Activity 13*.

Class Activity 12: Slides, Turns and Flips

The laws of nature are but the mathematical thoughts of God.

Euclid

There are three basic rigid motions of the plane – ways to move the plane without distorting it. You may have learned about them informally in middle or high school or perhaps in an earlier course. Here you will study the precise definitions for those rigid motions and you will use those definitions to figure out how to construct each rigid motion.

A **translation** by a vector *RS* is a motion of the plane so that if *A* is any point in the plane and we call *A'* the image of *A*, then vector *AA'* and vector *RS* have the same length and direction. We will denote this translation T_{RS}.

1) Construct △*A'B'C'* (the image of △*ABC* under the translation T_{RS}) and then prove, *using the definition of a translation*, that you have done so.

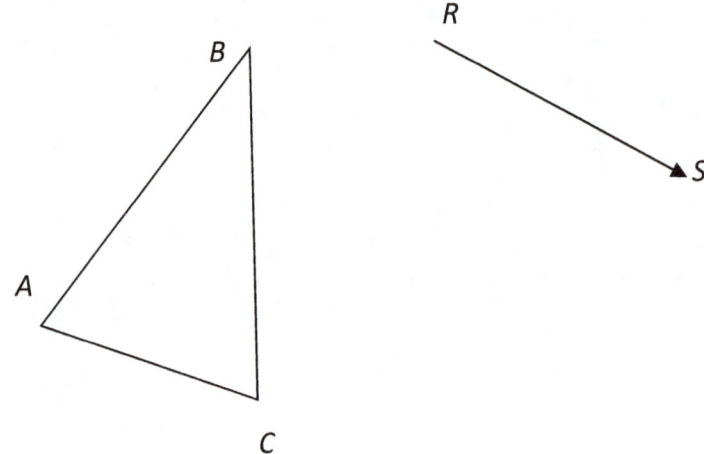

(This activity is continued on the next page.)

81

A **rotation** (about center point P of angle φ) is a motion of the plane in which the image of P is itself, and if the image of A is A' then PA' is congruent to PA and the measure of angle APA' = φ. We will denote this rotation R$_{(P, φ)}$.

2) Construct ΔA'B'C' (the image of ΔABC under the clockwise rotation R$_{(P, φ)}$) and then prove, *using the definition of a rotation*, that you have done so.

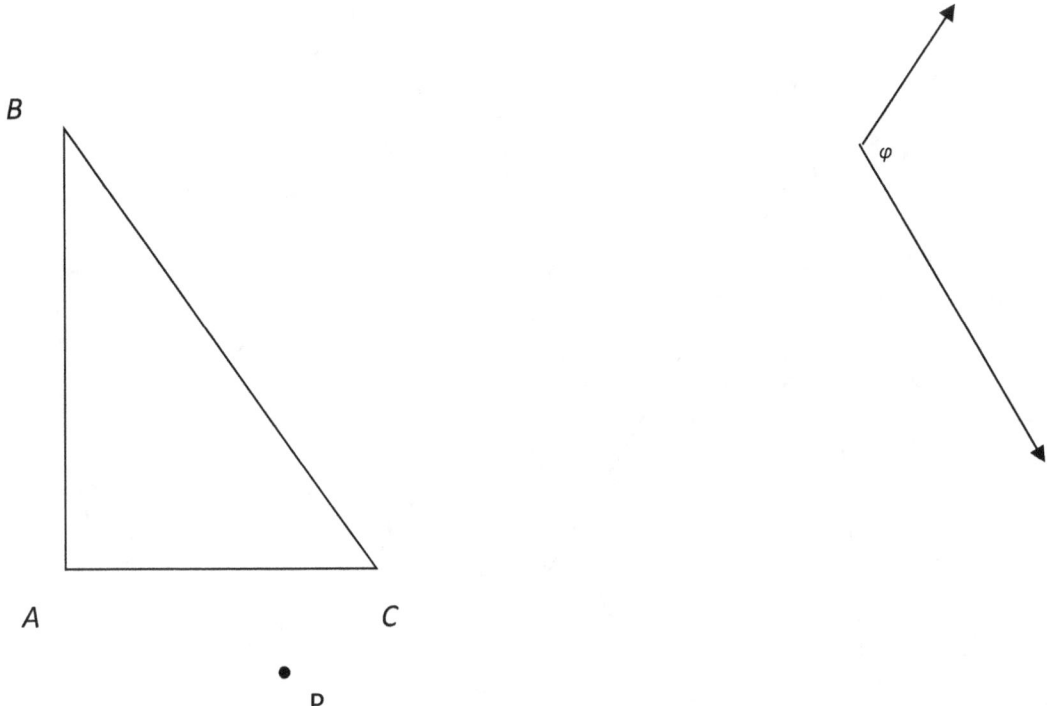

(This activity is continued on the next page.)

A **reflection** (in line *m*) is a motion of the plane in which the image of a point on *m* is itself, and if *A* is not on *m* and *A'* is the image of *A*, then *m* is the perpendicular bisector of *AA'*. We will denote this reflection M$_m$. (M for mirror.)

3) Construct △*A'B'C'* (the image of △*ABC* under the reflection M$_m$) and then prove, *using the definition of a reflection*, that you have done so.

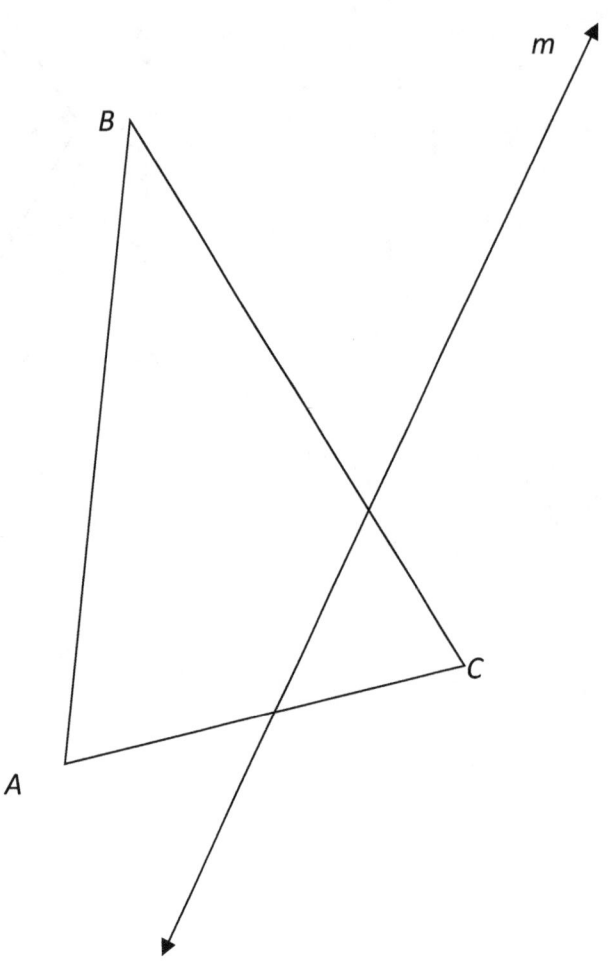

Read and Study:

> *Say what you know, do what you must, come what may.*
> *Sonja Kovalevsky (Motto on her paper "On the Problem of the Rotation of a Solid Body about a Fixed Point.")*

Informally, **transformations** refers to a large category of motions that can be applied to the entire plane. Transformations can tear, stretch, compress, or simply move the plane, potentially changing an object's shape, size, orientation, and position.

Rigid motions (also known as **isometries**) are a special class (subset) of transformations that do *not* cause distortions. You can think of them like this: suppose that you set an infinite piece of paper on the table in front of you and picture that paper as representing the set of points in the plane. Now, what can you do to move this paper so that in the end it is back flat on the table? Well, you could spin it around (i.e., perform a rotation); you could flip it over (i.e., perform a reflection); you could slide it in some direction (i.e., perform a translation); or you could do some combination of these moves. If you stretch the paper, crumple it, or tear it, then you have *not* performed a rigid motion. A dilation is an example of a motion of the plane that is *not* rigid – informally, a **dilation** is a stretching or a shrinking of the plane (and all of the objects on the plane) in a uniform manner.

Here are the more technical definitions:

- A **transformation** on a set S is a function from S back to S that is both one-to-one and onto.

- An **isometry** (**rigid motion**) on a set S is a transformation of the set S that preserves distances, meaning that the distance between any points P, Q is equal to the distance between their images P', Q'

The requirement that isometries preserve distances is very strong. It guarantees that objects are congruent to their images, ensuring the 'no distortion' rule.

So that you understand these definitions, let's first look at transformations in general, beginning with what we mean by 'function', 'one-to-one', and 'onto.'

A **function** (from set S to set T) is a rule that assigns to each element of the set S an element of set T in such a way that every element in S is paired with an element of T and no element of S is assigned to more than one element of T. If S and T were groups of people, this would be like saying that everyone in S has exactly one buddy in T.

A function is **one-to-one** if no element of T is assigned to more than one element of S. Using our buddy example, "one-to-one" is like saying that everyone in S has a *different* buddy in T (i.e. no two people in S have the same buddy in T).

A function is **onto** if every element of T is assigned to some element from S. Using our people example, "onto" is like saying everyone in T has a buddy in S (i.e. no one in T is left out).

Okay, now stop and think about all this.

1. *Give an example of a function from the set S = {a, b, c, d} to the set T = {2, 4, 6} that is onto but not one-to-one. Think through that definition. Remember that in mathematics understanding definitions is the key to understanding ideas.*

2. *Give an example of a function from the set S = {a, b, c} to the set T = {2, 4, 6, 8} that is one-to-one but not onto.*

3. *Give examples of functions that are both one-to-one and onto. (You get to make up the sets too.)*

Really try to do all the above before you turn this page!

Okay, here are our examples.

1) $a \to 4, b \to 2, c \to 6, d \to 2$

   Notice that this is a function from S to T because every element of S is paired with an element of T and no element of S is paired with more than one element of T. It is also "onto" because everyone in T is assigned to something from S, but it is not "one-to-one" because b and d share a buddy in T.

2) $a \to 6, b \to 8, c \to 4$

   This function is "one-to-one" because no element of T is assigned to more than one element of S. However it is not "onto" because 2 is an element of T with no buddy in S assigned to it.

3) 1st example: Let A = {f, d} and B = {x, v} then the assignments $f \to v$ and $d \to x$ is a function that is both one-to-one and onto. *Explain why.*
   2nd example: Let S and T be the same set $\mathbb{R}$ (i.e. the set of real numbers). Then we can define a transformation ϕ (of the set $\mathbb{R}$) by taking any element x of $\mathbb{R}$ to x + 3. *Explain why this function is both one-to-one and onto and a transformation.* We usually denote that function like this: $\phi(x) = x + 3$.
   3rd example: Let f be the function from $\mathbb{R}$ to $\mathbb{R}$ defined by: $f(x) = x^2$. *Is f one-to-one? Onto? A transformation? Explain.*

Based on this, a transformation of the plane is a function from the plane to itself that is one-to-one and onto. It is important to note that transformations describe a motion of the *entire plane* – not just a few objects in the plane. For example, when you used the definition of a reflection to create $\Delta A'B'C'$, you did not create a reflection; the triangle you created (often called the triangle's **image**) just showed where the original triangle would be found if you reflected the *entire plane* in line *m*. In other words, our construction of the triangle's image gives us an incomplete view of where part of the plane goes when it is reflected in line *m*. *This is a very important idea, so make sure you understand it.*

All of the motions you explored in the class activity (translations, rotations, and reflections) are transformations of the plane. *Take some time to revisit the definition of each one and explain why it is a transformation of the plane.*

A point P is a **fixed point** of a transformation, if the image of P is P itself. Makes sense right? Fixed points are those that do not "move" under the transformation, or, said another way, fixed points are the points that get paired with *themselves* under the function. *Revisit the definitions of translations, rotations, and reflections. What fixed points does each have? How do you know?*

So far, we have been talking about transformations in general. Let's end by looking at something that's special about isometries (rigid motions). *Go back and re-read our technical definition of an isometry. What makes isometries different from other transformations?*

The figure below depicts two triangles ($\triangle ABC$ and $\triangle A'B'C'$) along with three additional points (E, F, and G). The points A', B', and C' are the image of points A, B, and C (respectively) under some isometry of the plane. Using **only the definition of isometry** and your construction tools, determine where the images of E, F, and G (i.e. E', F', and G') would be found.

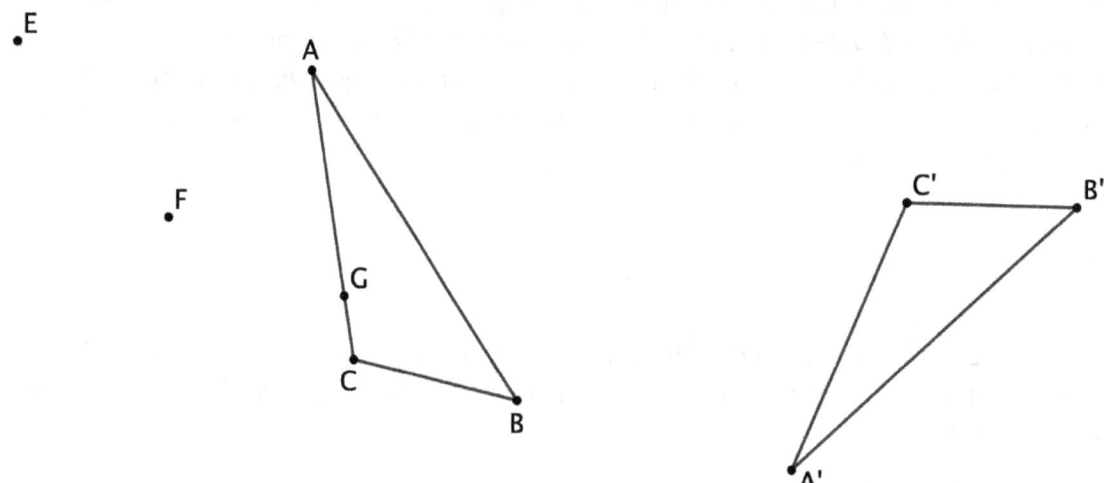

Before you turn the page, complete the task and ponder: How were you able to do it without knowing more about the isometry? How do you know you are correct?

The distance-preserving requirement used to define isometries is very powerful. Because of it, we get two amazing results:

- If we know what an isometry does to three non-collinear points, then we can determine what it does to *every* point in the plane.

- If two isometries agree on three non-collinear points, then they are the same isometry. More specifically, we would say that given isometries f and g, if $f(P) = g(P)$ for three non-collinear points P, then $f(P) = g(P)$ for *every* point P in the entire plane.

In other words, an isometry is completely determined by what it does to a single triangle (or three non-collinear points). This is why we so often use triangles and their images to represent the effects of an isometry.

Homework:

With regard to excellence, it is not enough to know, but we must try to have and use it.
Aristotle

1) Do all the italicized things in the *Read and Study* and *Connections* sections.
1) Consider the functions from $S = \{a, b, c\}$ to S below.

$$
\begin{array}{cc}
f & g \\
a \to b & a \to c \\
b \to c & b \to b \\
c \to a & c \to a
\end{array}
$$

a) Is f one-to-one? Is g one-to-one? Explain.
b) Is f onto? Is g onto? Explain.
c) Which of these two functions satisfy the definition of a transformation?
d) Make an arrow chart like we did above defining each of the following functions:
$f \circ g$, $g \circ f$, $f \circ f$ and $g \circ g$.
e) Does f have any fixed points? Does g? Does $f \circ g$ have any fixed points?

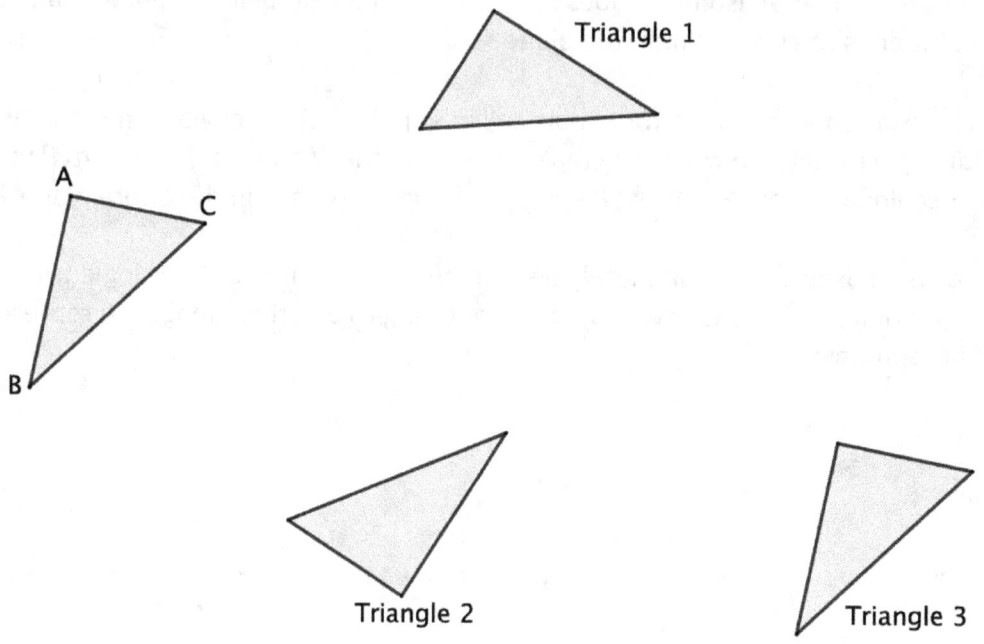

2) Above is a representation of a triangle *ABC* and three additional triangles, numbered 1 – 3.
   a) Which of the three triangles *could* and which *could not* be the **translated** image of triangle *ABC*?
   b) Explain how you know in each case.
   c) For those that could, how could you *construct* the translation vector?
3) Redo #2, but for a rotation. In other words:
   a) Which of the three triangles *could* and which *could not* be the **rotated** image of triangle *ABC*?
   b) Explain how you know in each case.
   c) For those that could, how could you *construct* the center and angle of rotation?
4) Redo #3, but for a reflection. In other words:
   a) Which of the three triangles *could* and which *could not* be the **reflected** image of triangle *ABC*?
   b) Explain how you know in each case.
   c) For those that could, how could you *construct* the line of reflection?

5) Below is a representation of two congruent triangles.
    a) Could one of the triangles below be the image of the other under a translation?
    b) What about under a rotation?
    c) What about under a reflection?
    d) Give your best argument as to why you are correct in each case.

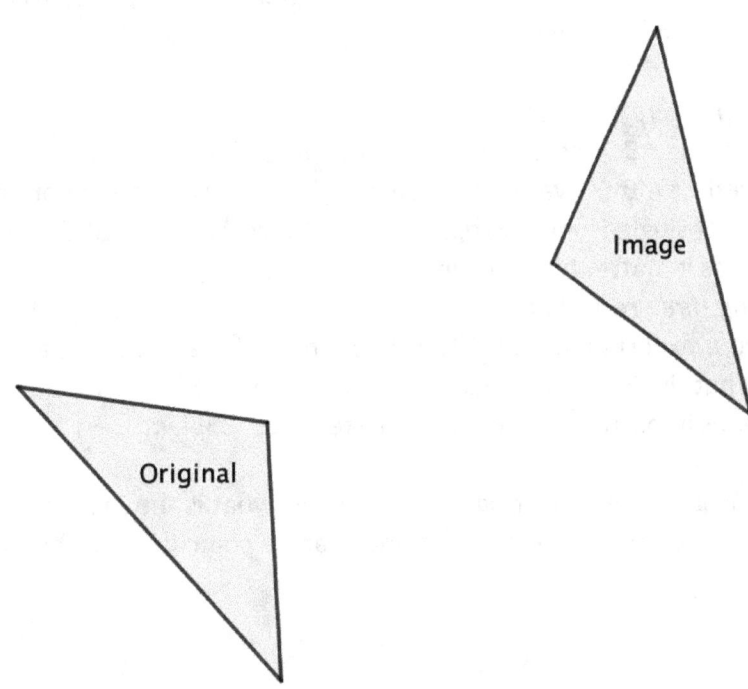

6) Given an object and its image, explain how you can construct the following (assuming they exist)? In each case, explain.
    a) The translation vector
    b) The center and angle of rotation
    c) The line of reflection
7) How many fixed points does each type of isometry have? In each case, explain.
    a) Translation
    b) Rotation
    c) Reflection

Class Activity 13: Flip-Flop

Without reflection, we go blindly on our way, creating more unintended consequences, and failing to achieve anything useful.

Margaret J. Wheatley

In this activity you will explore the composition of reflections (i.e. the result of performing two reflections, one following the other). The first reflection will be applied to the original object. The second will be applied to the image of the first.

1) *Composing two reflections*:
   Draw a triangle *ABC* and draw two lines of reflection *m* and *n*. First, reflect *ABC* over line *m*, then reflect *A'B'C'* over *n*. In each case, carefully *construct* your reflection.
   a. What single isometry (rigid motion) takes *ABC* to *A"B"C"*? How do you know?
   b. Does it matter how the lines are related? Explain.
2) *Composing three reflections*:
   Repeat your experiment, but with reflections in three lines *l*, *m*, and *n*.
   a. What single transformation takes *ABC* to *A'''B'''C'''*? How do you know?
   b. Does it matter how the lines are related? Explain.
3) *Exploration*:
   If someone gave you two congruent triangles, what is the most/least number of reflections it would take to make the image of one triangle coincide with the other triangle?

Read and Study:

If I have ever made any valuable discoveries, it has been owing more to patient attention, than to any other talent.
                                                                    Sir Isaac Newton

In the class activity, you began working with composing isometries. Speaking of functions in general, the **composition** of two functions is an operation that takes two functions and creates a new one. It is defined as follows: *f* **composed with** *g* (written $f \circ g$) is the function whose values are given by $f \circ g(P) = f(g(P))$. In other words, the value of $f \circ g$ comes from doing *g* first, and then putting the output of *g* through *f*.

Okay, let's think about this more carefully. Notice that for the composition $f \circ g$ to make sense, the outputs (range) of *g* must also be inputs (in the domain) of *f*. This can cause complications when working with arbitrary functions; however, when working with transformations there will be no problem performing compositions, because the inputs and outputs are in the same set. *Explain why this is the case.*

Because isometries are transformations, we can compose them by performing one after the other. In that case we say you have performed a **composition** of isometries. For example, you could do a rotation followed by a reflection. Or a translation followed by another translation. You can also compose rigid motions with dilations.

In the class activity, we asked you to compose reflections and then asked you to determine what single isometry the composition gave. Underlying this task is are two important theorems:

Theorem 1: The composition of two transformations is a transformation; furthermore, if they are both isometries, then the composition is also an isometry.

Look up the definitions for both and write your best argument for why this is the case.

This theorem gives us a nice way of defining two important ideas: congruence and similarity. We say that two geometric objects are **congruent** if one is the image of the other under an isometry (i.e. rigid motion). This automatically includes compositions of isometries, because the result of composing isometries is still an isometry.

We say that two geometric objects are **similar** if one is the image of the other under a composition of an isometry and a dilation. (In other words, objects are similar if one can moved and shrunk (or made larger (dilated)) so as to coincide with the other.) Similar objects are the shape but not necessarily the same size. For example, these snowflakes are similar but not congruent. We will discuss this idea further in an upcoming section.

In the previous section, we looked at three different types of isometries: translations, rotations, and reflections. As you composed reflections, however, you should have discovered that these three types do not include all isometries. By composing three reflections, it is possible (and highly likely) to get an isometry that is neither a translation, nor rotation, nor reflection.

This new type of isometry is called a **glide reflection**, also known as a "slide flip". It is usually defined as the composition of a translation and reflection, where the translation vector and line of reflection are parallel. We will denote a glide reflection with translation vector RS and line of reflection m by $G_{(RS,m)}$ (G for glide reflection.) A common example of a glide reflection is a set of foot prints in the sand. You'll play around with glide reflections more in the homework.

Another isometry you might have encountered is the **identity** transformation (usually denoted by the Greek letter ι (iota)). Formally, for any set S and point P in that set, the identity is the transformation ι where $\iota(P) = P$. Why is this an isometry? Write your best argument.

You might be wondering whether there are even more types of isometries that we haven't brought up—or even discovered. The short answer is no; every isometry can be classified as a translation, rotation, reflection, or glide reflection. *How might we classify the identity transformation using these?*

At this point, if you are a good skeptic, you should be wondering, "How do we know we have them all?" The answer lies in your work during the class activity. Let's put together what we know with what you uncovered:
- First, (from last section) every isometry is completely determined by what it does to a triangle.
- Second, (from #3 in the class activity) any isometry can be written as the composition of at most three reflections, since any triangle can be superimposed on its image by the composition of at most three reflections.
- Third, an isometry can be recreated using as few as:
    - 0 reflections, if it's ι (the identity transformation).
    - 1 reflection, if it's a reflection.
    - 2 reflections, if it's a translation or a rotation.
    - 3 reflections, if it's a glide reflection.

We have just argued a major theorem about isometries, namely:

Theorem 2: Every isometry of the plane is a translation, rotation, reflection, or glide reflection.

That's it! So, there are only four types of isometries of the plane, meaning that we have a complete classification of all isometries.

Another interesting idea that comes from reflections is what it does to rotational orientation. Take any triangle *ABC*. We will say that triangle *ABC* is **oriented clockwise** if going from *A* to *B* to *C* to *A* involves moving clockwise around the interior; otherwise, we will say it is **oriented counter-clockwise**.

Draw a triangle ABC and label it so that it is oriented clockwise. Now construct its image (triangle A'B'C') under a translation. Does the image have the same rotational orientation (clockwise) as the original triangle or did it change (i.e. to counter-clockwise)? Will this same thing happen for every translation? Explain.

Repeat this experiment, but using a reflection instead. What happened? Explain.

What will happen if you do this with a rotation or glide reflection? Try it and find out.

Some properties are useful in distinguishing among the different types of isometries. Fixed points is one such example. How an isometry affects the rotational orientation of objects and their images is another. *Except for the last row, complete the table below with what you have learned about translations, rotations, and reflections, each type of rigid motion by describing what fixed points it has and telling whether objects and their images have the same or opposite (clockwise/counter-clockwise) orientations:*

	Fixed Points	CW/CCW Orientation of Objects & their Images
Identity Isometry (ι)	Every point in the plane	Unchanged
Translation (T_{RS})		
Rotation ($R_{(P, \varphi)}$)		
Reflection (M_m)		
Glide Reflection ($G_{(RS, l)}$)		

One last question for you to ponder: Is there a relationship between the last column and how that isometry can be created using the composition of reflections?

Connections to Teaching:

In grades 6-8 all students should describe sizes, positions, and orientations of shapes under informal transformations such as flips, turns, slides, and scaling.

National Council of Teachers of Mathematics
Principles and Standards for School Mathematics, p. 232

The Common Core State Standards recommend that students in Grade 8 learn to do and understand the following. *Read these carefully.*

Geometry Grade Eight:
Understand congruence and similarity using physical models, transparencies, or geometry software.
1. Verify experimentally the properties of rotations, reflections, and translations:
    a. Lines are taken to lines, and line segments to line segments of the same length.
    b. Angles are taken to angles of the same measure.
    c. Parallel lines are taken to parallel lines.

2. Understand that a two-dimensional figure is congruent to another if the second can be obtained from the first by a sequence of rotations, reflections, and translations; given two congruent figures, describe a sequence that exhibits the congruence between them.

3. Describe the effect of dilations, translations, rotations, and reflections on two-dimensional figures using coordinates.

4. Understand that a two-dimensional figure is similar to another if the second can be obtained from the first by a sequence of rotations, reflections, translations, and dilations; given two similar two-dimensional figures, describe a sequence that exhibits the similarity between them.

Copyright 2010. National Governors Association Center for Best Practices and Council of Chief State School Officers. All rights reserved.

Middle grades students will likely not *construct* rigid motions; rather they will use graph paper or tracing paper to study them informally. *To see what we mean, get a few pieces of tracing paper and re-do the class activity by tracing the figures and moving your paper.*

Homework:

With regard to excellence, it is not enough to know, but we must try to have and use it.
Aristotle

- Do all the italicized things in the *Read and Study* and *Connections* sections.

- Let's begin unpacking the definition of a **glide reflection**. Study the definition as you do the following:
    a) Use the grid to perform $G_{RS,\overrightarrow{RS}}$ as $M_{\overleftrightarrow{RS}} \circ T_{RS}$ by: first, translating *ABCD* by the translation vector *RS*, then reflecting *A'B'C'D'* over line *RS*.
    (You do *not* need to construct these rigid motions.)

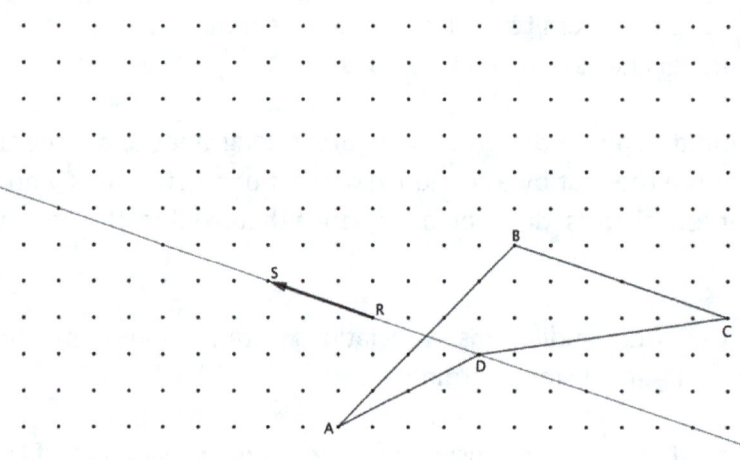

    b) Now, perform $G_{RS,\overrightarrow{RS}}$ as $T_{RS} \circ M_{\overleftrightarrow{RS}}$ by the same glide reflection by: first, reflecting *ABCD* over line *RS*, then, translating *A'B'C'D'* by the translation vector *RS*.
    c) Does it matter whether the translation or reflection comes first? Explain.
    d) Study your work and complete the last row of the table in the *Read and Study*, by describing glide reflections' fixed points and impact on cw/ccw orientations.

- The figure below depicts a quadrilateral and its image under an unspecified glide reflection. Determine a way to reconstruct the line of reflection and translation vector for the glide reflection—remember that the vector and line need to be parallel. It may be useful to revisit the last problem and see how they are related to the line segments that join corresponding vertices.

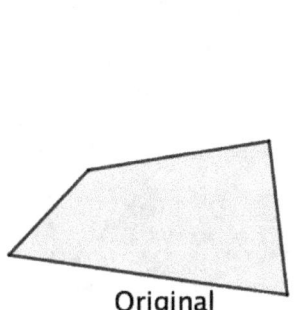
Original

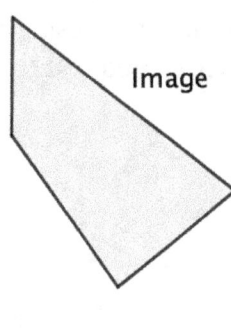
Image

- Explore the composition of two translations by doing the following with the figure provided below:
    a) Perform the composition $T_{ST} \circ T_{RS}$ by: first, translating ABCD by the vector RS, then translating A'B'C'D' by the vector ST.
    b) Describe the single isometry that would equal the composition.
    c) Now, perform the composition $T_{RS} \circ T_{ST}$ by: first, translating ABCD by the vector ST, then translating A'B'C'D' by the vector RS.
    d) When composing two translations, does it matter which is performed first?

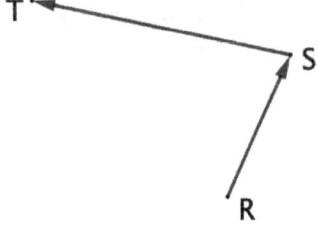

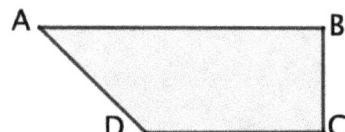

- Explore the composition of two rotations (with the same center) by doing the following:
    a) Construct the composition $R_{(P,90°)} \circ R_{(P,60°)}$ by: first, rotating ABCD clockwise around P by 60°, then rotating A'B'C'D' clockwise around P by 90°.
    b) Describe the single isometry that would equal the composition.
    c) Now, construct the composition $R_{(P,60°)} \circ R_{(P,90°)}$ by by: first, rotating ABCD clockwise around P by 90°, then rotating A'B'C'D' clockwise around P by 60°. .
    d) When composing two rotations with the same center, does it matter which is performed first?

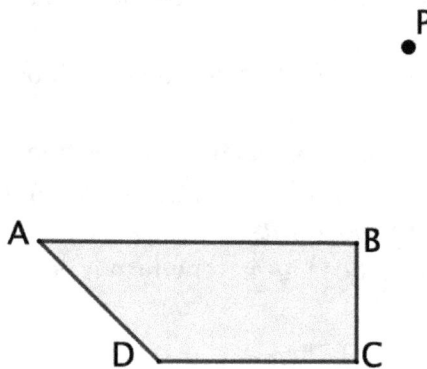

- In the Read & Study, we said that every isometry is the composition of at most three reflections. Determine whether such a composition is unique. In other words, is there only one way to write an isomorphism as the composition of reflections?

99

Class Activity 14: Transformative Thinking

The mathematical sciences particularly exhibit order, symmetry, and limitation; and these are the greatest forms of the beautiful.
                                                                                            Aristotle

In the last section, you investigated the composition of reflections and learned that there are only four types of isometries. In this activity, your job is to classify all possible isometries of the plane.

∘	Translation T_{RS}	Rotation $R_{(P,\varphi)}$		Reflection M_l		Glide Reflection $G_{(RS,l)}$	
Translation T_{PQ}						$PQ \parallel RS$	$PQ \nparallel RS$
Rotation $R_{(Q,\vartheta)}$		$P = Q$	$P \neq Q$				
Reflection M_n				$n \parallel l$	$n \nparallel l$, $n \neq l$	$n \parallel l$	$n \nparallel l$, $n \neq l$
Glide Reflection $G_{(PQ,n)}$						$n \parallel l$	$n \nparallel l$, $n \neq l$

100

Read and Study:

> *If I have ever made any valuable discoveries, it has been owing more to patient attention, than to any other talent.*
> Sir Isaac Newton

In the last section, we discussed composition of functions. Composition has some really important properties for transformations and isometries. Let's discuss some of them.

First, composition is associative. This means that if $\alpha, \beta,$ and γ are transformations, then $(\alpha \circ \beta) \circ \gamma = \alpha \circ (\beta \circ \gamma)$, which means that when composing a sequence of transformations, as long as we keep them in the same order, it does *not* matter which composition operation is performed first, the end result is the same transformation.

Second, if α (alpha) and β (beta) are any two transformations on a set S, then $\alpha \circ \beta$ is a transformation on S. That means that transformations are **closed** under composition. Similarly, isometries are closed under composition. *Explain what this means and why it is true.*

Third, there is an identity transformation (also an isometry). Remember the identity transformation ι (iota), where $\iota(P) = P$, for every point P? When we compose the identity transformation with any other transformation α on S, we have that $\alpha \circ \iota = \iota \circ \alpha = \alpha$. This equation means that composing ι with another transformation is the same as just doing that other transformation.

Fourth, every transformation (or isometry) has an inverse transformation (or isometry). We say that a transformation β is the **inverse** of a transformation α if and only if $\alpha \circ \beta = \beta \circ \alpha = \iota$, and we denote the inverse of α as α^{-1}. We like to think of the inverse transformation as the transformation that undoes the transformation. *Find the inverse of each of these isometries:*
- A translation by vector RS

- A rotation clockwise 55° about a point P

- A reflection over line m

- A glide reflection with translation vector RS and line of reflection m

A set **T** of transformations on a set S is a **transformation group** if and only if the following properties hold:
   a) (closure) The composition of two transformations in **T** is in **T**.
   b) (identity) The identity transformation, ι, is in **T**.
   c) (inverses) If a transformation α is in **T**, then its inverse, α^{-1}, is also in **T**.

Note: A transformation group is ***not*** an Abelian group. That is, the operation of composition of transformations is not commutative. Recall that an operation, *, is commutative when $x * y = y * x$ for all elements in the set. *Give an example of two transformations α and β where $\alpha \circ \beta \neq \beta \circ \alpha$.*

Putting all this information together, we get this important result:

Theorem: The set of all isometries is a transformation group.

Look back up at all the definitions. Explain carefully what this theorem means.

We know that we have thrown a lot of definitions at you and things may be a little confusing right now. We want you to focus on understanding what they say. This will be your main job in the homework for this section.

Homework:

In this life, we get only those things for which we hunt, for which we strive, and for which we are willing to sacrifice.
George Matthew Adams

1) Do all the italicized things in the *Read and Study* section.

2) In the readings, we said that the set **T** of all transformations of the plane forms a transformational group. Does the set **V** of all translations of the plane form a transformational group? If so, explain how you know it meets the definition. If not, explain why not.

3) Does the set **R** of all rotations of the plane form a transformational group? If so, explain how you know it meets the definition. If not, explain why not.

4) Does the set **P** of all rotations about point *P* form a transformational group? If so, explain how you know it meets the definition. If not, explain why not. (Does this change your answer to the previous question?)

5) Does the set **M** of all reflections of the plane form a transformational group? If so, explain how you know it meets the definition. If not, explain why not.

6) Does the set **G** of all glide reflections of the plane form a transformational group? If so, explain how you know it meets the definition. If not, explain why not.

Class Activity 15: Tessellations

The mathematician's patterns, like the painter's or the poet's must be beautiful; the ideas, like the colors or the words must fit together in a harmonious way. Beauty is the first test: there is no permanent place in this world for ugly mathematics.

Godfrey H. Hardy

There are many new definitions involved in this activity. Take a little time to study each of them.

A **tiling** is an arrangement of polygons that can be extended in all directions to cover the plane with no gaps and no overlaps.

A **tessellation** is a tiling in which all vertices meet only other vertices.

A **regular tessellation** is a tessellation that uses only *one* regular polygon.

Find all the possible regular tessellations, make a sketch of each one, and then make an argument that you have found them all.

Here are traceable copies of many regular polygons. All of them have been scaled so that they have the same side length.

Read and Study

Everything has beauty, but not everyone sees it.

Confucius

A **tiling** is an arrangement of polygons that can be extended in all directions to cover the plane with no gaps and no overlaps. An example using a "T" shape polygon is shown below.

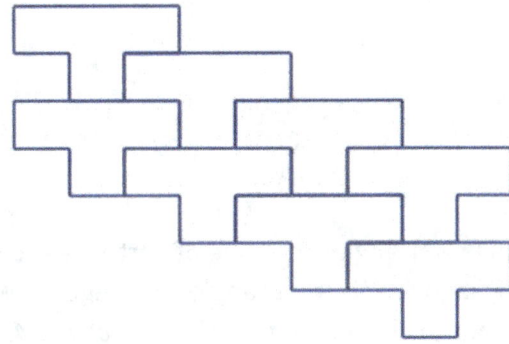

A **tessellation** is a tiling in which all vertices only meet other vertices. *Is the above tiling a tessellation? Explain.*

Some curriculum materials use the words tiling and tessellation interchangeably. We will not do so, but we want you to be aware of that fact.

Mathematically, we are interested in interpreting the definition of a tessellation. How can we know it will have no gaps and no overlaps? How can we determine that a given arrangement will extend indefinitely in all directions? First we will examine what happens at a single vertex point within a tessellation.

Take a close look at vertices A and B in the following arrangement composed of regular hexagons and equilateral triangles.

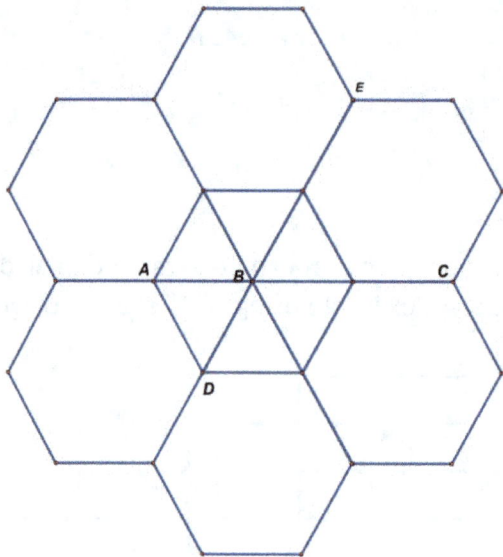

There are two hexagons and two triangles meeting at vertex A and six triangles meeting at vertex B. Since we know that the hexagons and the triangles are regular, we know that the vertex angles of the hexagon are 120° each. (*Have a look back at Class Activity 4, if you have forgotten how to think about this.*) Likewise, we know that the vertex angles of the equilateral triangle are 60° each. This means that there are 2*120° + 2* 60° = 360° at vertex A and there are 6*60° = 360° at vertex B. (*Check this.*) Now, since the sum of the angles at each vertex is exactly 360°, we have proved that there are no gaps or no overlaps in this arrangement.

Now we will note that if we made infinitely many copies of this arrangement, we could slide (translate) them around on the plane (e.g., slide one so that A goes to C) to cover the entire plane.

We *will* tell you that there are many tessellations of regular polygons that are not regular. *For example, have a look back at the tessellation made of hexagons and triangles that we were just discussing above. Why is this tessellation not regular?*

Connections to the Elementary Grades

Learning is not compulsory... neither is survival.

W. Edwards Deming

Creating tessellations is an activity that can be adapted to every grade level – very young children can create patchwork quilts from construction paper using only rectangles or squares or isosceles right triangles. Older students can make more complex artwork using the mathematical concepts of congruency and transformations.

Online resources can allow students to create a variety of more complicated designs quickly and accurately. *Spend 15 minutes at the website at* http://www.mathcats.com/explore/tessellations/tesspeople.html *to see an example of an interactive web-based exploration of tessellations used in the fifth grade Trailblazers curriculum in North Carolina.*

Homework

Creativity is allowing yourself to make mistakes. Art is knowing which ones to keep.
Scott Adams, 'The Dilbert Principle'

1. Do all the italicized things in the *Read and Study* and *Connections* sections. Make sure to spend some time at the website.

2. Any triangle (if you have enough copies of it) can be used to tessellate the plane. To explore this, fold a paper up so you can cut out 8 congruent **scalene** triangles all at once. Then have a look. Pay particular attention to the transformations you are using to move the triangle into new positions.

    a. Make a mathematical argument that your triangle would, in fact, tessellate the plane.
    b. Where in your proof did you use transformations of the plane?

3. True or false? Any quadrilateral (if you have enough copies of it) can be used to tessellate the plane. To explore this, fold a paper up so you can cut out 8 congruent quadrilaterals all at once. Then have a look. What about concave quadrilaterals? Make an argument to support your choice.

4. True or false? It's possible to find a pentagon that can be used to tessellate the plane (if you have enough copies of it). Make an argument to support your choice.

5. Make a mathematical argument that the number of regular tessellations you found in the class activity is the exact number possible.

6. A polygon with more than six sides will not tile if it is convex. Explain why not. The following polygons have more than six sides, but they are concave. Sketch a portion of a tiling for each polygon.

a)   b)

7. Describe some reflectional, rotational and translational symmetries for the tessellation depicted below. Remember that tessellations continue forever to cover the plane.

Class Activity 16: Expanding and Contracting

Give me extension and motion and I will construct the universe.
Rene Descartes

Another motion of the plane is a dilation. You have experienced dilations whenever you shrink or enlarge a photograph without distorting the image. Formally, a **dilation** (with center P and scale factor $k > 0$) is a motion of the plane in which the image of P is itself and the image A' of any other point A is on the ray PA, so that the distance PA' is k times the distance PA.

1) Unlike the rigid motions, dilations are not always constructible. (i.e. You cannot always make them with a compass and straight edge alone.) Why is this the case?

2) In the space below or on a separate sheet of paper, create a triangle $\triangle ABC$. Draw several dilations of $\triangle ABC$, experimenting with the location of the central point and the value of the scale factor; then answer the questions on the next page.

(This activity is continued on the next page.)

a. How does the placement of the center point affect the resulting shape?

b. How does the shape change if the scale factor is greater than one? Between 0 and 1? Equal to 1?

c. What would happen if we were to let $k = 0$?

d. Explain a reasonable way of thinking about a dilation if $k < 0$? How could you extend the definition to accommodate this idea?

Read and Study

In the physical world, one cannot increase the size or quantity of anything without changing its quality. Similar figures exist only in pure geometry.

Paul Valéry

Dilations are an example of a movement of the plane that is not a rigid motion. When you created your dilations in the class activity, you were creating shapes that were similar to the original figure. Recall that two figures are **similar** if one is the image of the other under a composition of rigid motions and dilations. For example, the following figures are similar. *See if you can determine a sequence of rigid motions and dilations which map one of the figures below onto the other.*

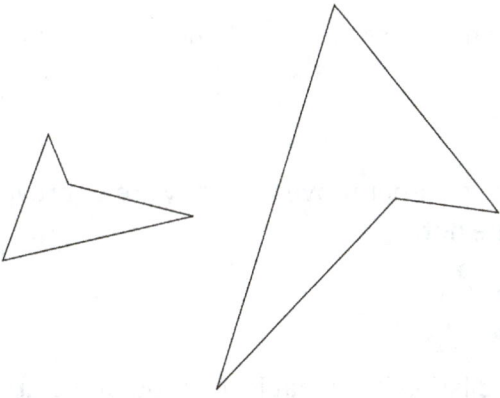

As a consequence of the definition of similar, we know that two polygons are similar if their corresponding vertex angles are congruent and corresponding sides are proportional. *Take a moment and think about why this is the case.*

Triangles are really special polygons in the fact that we do not have to check all the sides and all the angles to determine if two triangles are similar. We only have to check one of the following:

- Angle-Angle-Angle Similarity Theorem: If two triangles have corresponding angles congruent, then the triangles are similar. (This theorem is sometimes called the AA theorem because checking two angles is sufficient for proving that two triangles are similar. *Why is this the case?*)
- Side-Angle-Side Similarity Theorem: If two triangles have two pairs of corresponding sides proportional and the included angles congruent, then the triangles are similar.
- Side-Side-Side Similarity Theorem: If two triangles have all three pairs of corresponding sides proportional (with the same constant of proportionality), then the triangles are similar.

These theorems only work for triangles. *Why? What happens when you try to apply them to other polygons?*

Homework

A pupil from whom nothing is ever demanded which he cannot do, never does all he can.
John Stuart Mill

1) Do all the italicized things in the *Read and Study* section.

2) Determine if the following statements are true or false. Make sure you can explain why in each case.
    a. There are no fixed points in a dilation.
    b. Angles are preserved under a dilation.
    c. If two line segments are parallel before a dilation, they will be parallel after the dilation.
    d. Line segment lengths are preserved under a dilation.

3) Are all rectangles similar? Either prove that they are or provide a counter example explaining why they are not.

4) If the scale factor in a dilation is k, what is the ratio of the area of the resulting shape as compared to the original shape?

5) Determine if each pair of triangles below are similar. If they are similar, find the missing parts, if not, explain why not.

a.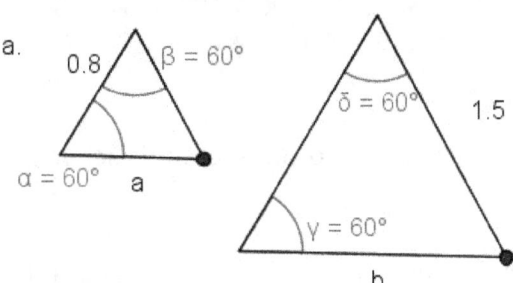

b.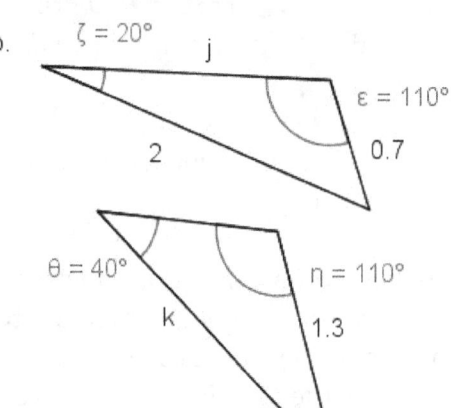

c.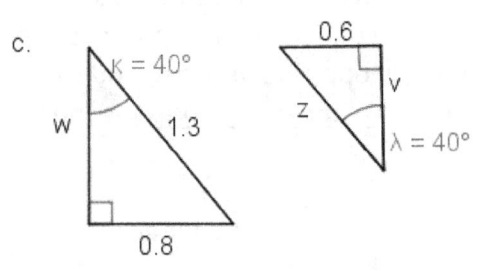

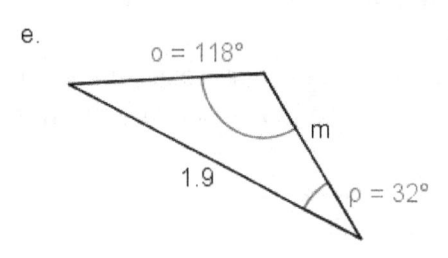

d.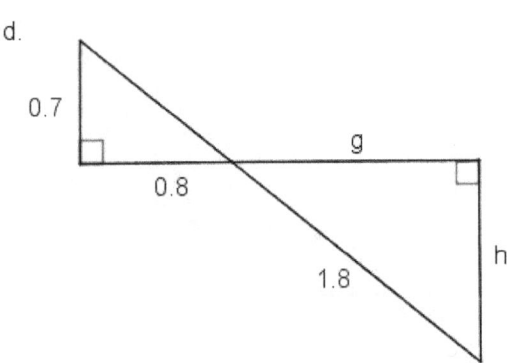

e.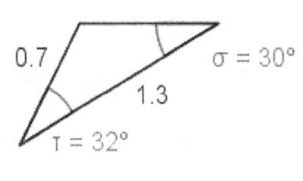

6) In the following figure assume that $\angle ACB$ is a right angle and $\overline{CD} \perp \overline{AB}$. Why are $\triangle ABC$, $\triangle ACD$, and $\triangle CBD$ all similar? Show that $cy = a^2$ and $cx = b^2$ and then use these facts to develop a careful proof of the Pythagorean Theorem.

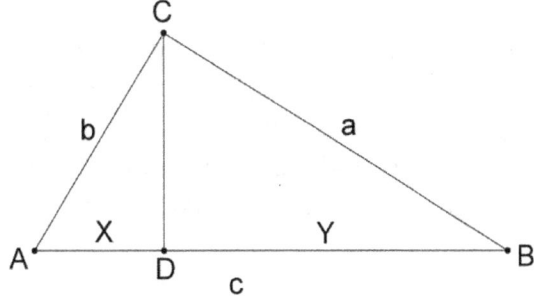

114

Class Activity 17: Strictly Platonic (Solids)

The most general law in nature is equity – the principle of balance and symmetry which guides the growth of forms along the lines of the greatest structural efficiency.
                                                                      Herbert Read

You will need to build models of the regular polyhedra in order to complete this activity – nets are available online.

Three-dimensional objects, including the regular polyhedra, can have rotational and reflectional symmetries. For rotational symmetry, the center of rotation is really a line of rotation (called the axis of symmetry). There can be more than one axis of symmetry for a three-dimensional object. For example, the cube has three axes of symmetry of order 4 connecting the centers of opposite faces, four axes of symmetry of order 3 connecting diagonally opposite vertices, and six axes of order 2 connecting midpoints of opposite edges. The **order of a line of symmetry** is the number of turns that put the object back on itself.

Here are the three order-4 axes of symmetry for a cube. Take a minute in your groups to be sure that everyone sees why each of these has order 4. Then sketch the rest of the axes of symmetry for a cube.

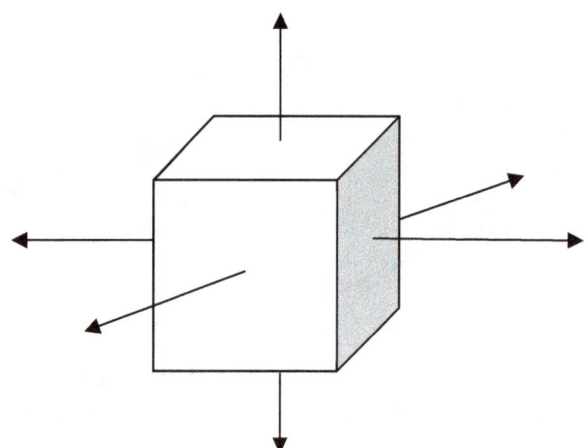

The cube also has reflectional symmetry. The line of reflection becomes a plane of reflection that divides the cube into two mirror images. There are nine planes of reflectional symmetry, two vertical, one horizontal and two through the diagonals of each pair of opposite faces. Find each plane of symmetry on your model of the cube. Imagine slicing your cube along each plane. You should be able to visualize the two congruent "half-cubes" that would result.

Your job for this *Class Activity* is to find and describe all the planes of reflectional symmetry and all the axes of rotational symmetry for the other four regular polyhedra. Complete the table and then describe any patterns you see.

Polyhedron	# and description of planes of reflection symmetry	# and description of lines of rotation symmetry
Regular Tetrahedron		
Cube		
Regular Octahedron		
Regular Dodecahedron		
Regular Icosahedron		

Read and Study:

The essence of mathematics is not to make simple things complicated, but to make complicated things simple.

S. Gudder

The geometric idea of symmetry is defined in terms of rigid motions. Here is the official definition.

A **symmetry** of a geometric object is a rigid motion of the plane in which the image of the object coincides with the original object.

Stop and think about this definition to be sure it makes sense to you.

Let's characterize an object in the plane based on its symmetries. Have a look at the two-sided arrow below.

This object has two reflection symmetries because reflections over either line shown below will result in the image coinciding exactly with the original object.

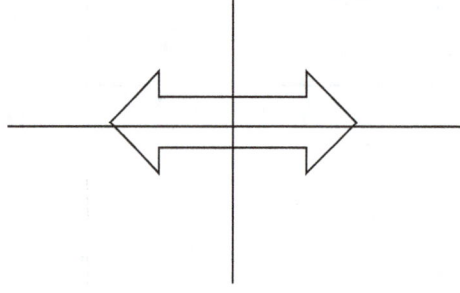

The two-sided arrow also has 180-degree rotational symmetry around the center (where the two lines above intersect). It also has 360-degree rotational symmetry (we call that the trivial symmetry because every object has it). *Draw an object that has 90, 180, 270 and 360 rotational symmetries and no other symmetries.*

What types of objects will have translational symmetries?

Have a look back at the table of symmetries you made for the regular polyhedra. *What do you notice?*

One thing that we noticed was that the cube and the octahedron have exactly the same set of symmetries, and that the dodecahedron and the icosahedron also share the same symmetries. So what is it about these pairs of objects that would have that be the case? Let's start with the cube and the octahedron. Imagine taking the midpoint of each face of the cube and thinking of those as the vertices of a new polyhedron. Then that new polyhedron would have 6 vertices. *See if you can sketch that new polyhedron inside of the cube.*

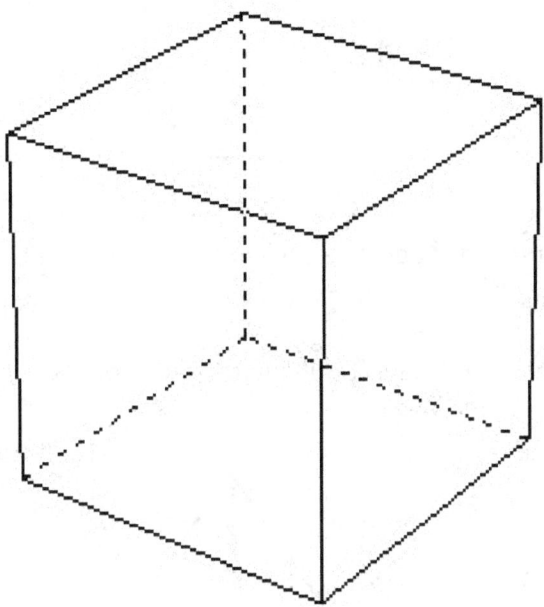

Now, see if you can sketch the polyhedron that could be formed by using the midpoints of the faces of the octahedron as vertices. *How many vertices would that new polyhedron have?*

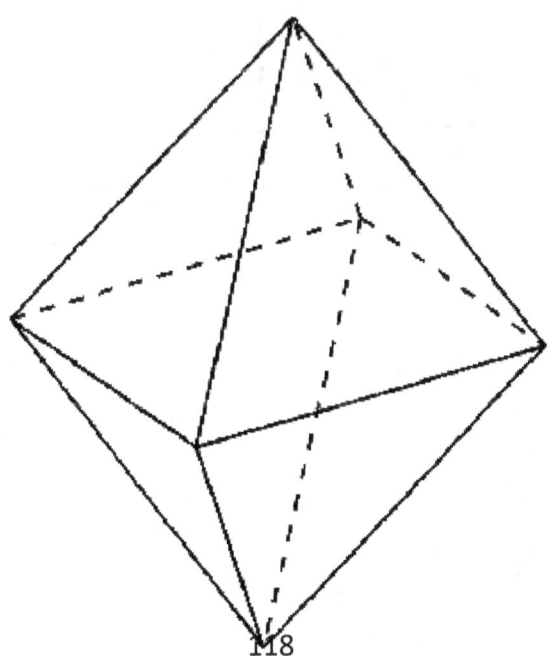

We call objects that are related in this way **duals**. The cube and the octahedron are duals of eachother, and the dodecahedron and the icosahedrton are also duals of eachother. *Take a close look at your models of the dodecahedron and the icosahedron to see if you can tell that they are duals.* We might even imagine how the objects would fit inside one another.

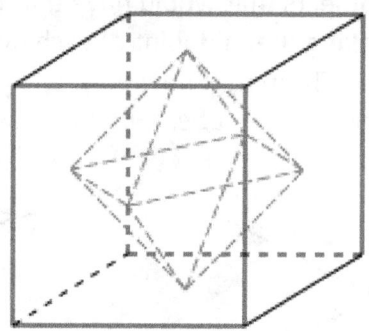

Duals have the same symmetries because they would move together under rotations and reflections.

You may have noticed that we have left out the tetrahedron. *What is its dual? See if you can sketch it.*

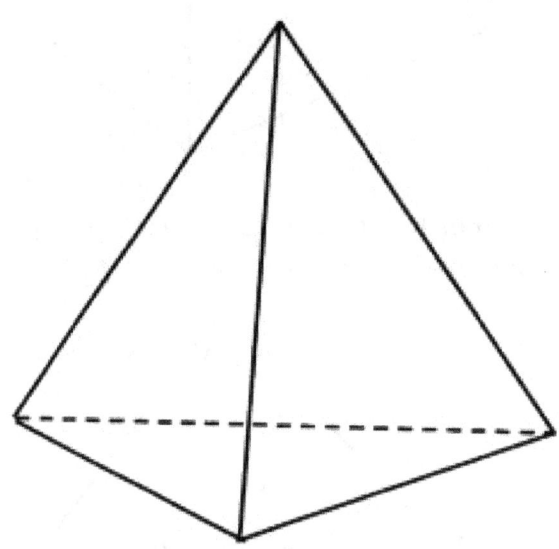

Homework:

You teach best what you most need to learn.

Richard Bach

1) Do all the italicized things in the *Read and Study* section.

2) Sketch an object *in the plane* that meets each set of criteria or explain why it is impossible to do so:
    a) The object has only 360-rotational symmetry.
    b) The object has 120, 240 and 360-degree rotational symmetries and no other symmetries.
    c) The object has 120 and 360-degree rotational symmetry and no other symmetries.
    d) The object has vertical reflection symmetry, 360-degree rotational symmetry and no other symmetries.
    e) The object has vertical translation symmetry, 360-degree rotational symmetry and no other symmetries.

3) An object in the plane has two lines of symmetry. If these lines are parallel, what other symmetries must this object have? Why?

4) Find all of the symmetries of the three-dimensional square-based pyramid shown below.

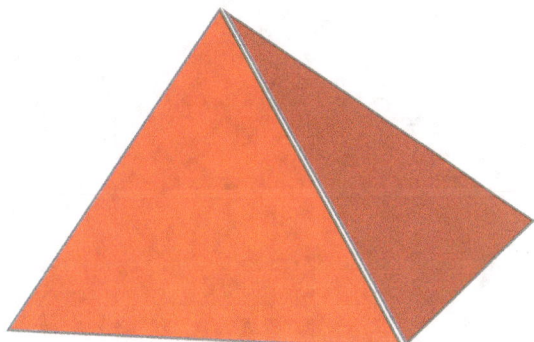

5) Prove that if an object in the plane has two intersecting lines of symmetry, then it must also have rotational symmetry.

6) Describe conditions which would guarantee that a right prism has exactly one plane of reflectional symmetry. Where is the plane located?

7) Describe conditions which would guarantee that an oblique prism has exactly one plane of reflectional symmetry. Where is the plane located?

8) Build the following models. Then find their surface areas, volumes, and describe all their symmetries.
   a) A right circular cylinder with radius 2 cm and height 7 cm.
   b) A square-based pyramid with a 4 cm by 4 cm base and a height of 5 cm.
   c) A right prism with the below regular hexagon as the base, and a height of 8 cm.

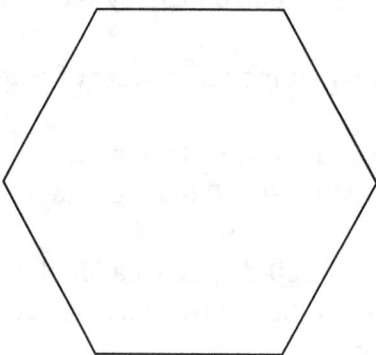

Class Activity 18: Buried Treasure

*I'm very well acquainted too with matters mathematical,
I understand equations, both the simple and quadratical,
About binomial theorem I'm teeming with a lot of news--
With many cheerful facts about the square of the hypotenuse.*

Gilbert & Sullivan, "The Pirates of Penzance"

The sneaky pirate and the first mate buried treasure on an island with two large rocks and palm tree near the shore. You've found the top-secret map that explains the location of the bounty as follows:

Me captain started at the palm tree and paced off the distance to the first rock, turned 90° in a counterclockwise direction and paced off an equal distance. Argh. I, the matey, started at the palm tree and paced off the distance to the second rock, then turned 90° in a clockwise direction and paced off an equal distance. We then buried the treasure halfway between us two.

You are standing on the island and the rocks are still there, but, sadly, the palm tree has long since died and you have no idea where it was. Find the treasure.

Read and Study:

Equations are just the boring part of mathematics. I attempt to see things in terms of geometry.

Stephen Hawking

In the 1700's René Descartes (pronounced Day-cart) had the idea that we could solve some geometric problems more easily by translating them into algebraic problems. His idea was to place a structure (a grid) on top of the Euclidean plane and to give names (like (-3, -1)) to the points. One version of the story goes like this: Descartes was not an early riser, but he enjoyed lying around in bed and thinking deeply. (Descartes is credited with the quote, "I think, therefore I am.") One morning, while pondering the ceiling of his bedroom, he noticed a fly walking across. As he mentally traced the path of the fly's walk he considered how he could describe the path mathematically. He reasoned that he could label any one point on the path by how far the fly was from the south wall and how far it was from the west wall of his room. Thus was born the idea of the coordinate plane upon which we can "see" the "path" of a function's graph.

The **coordinate plane** (also called the **Cartesian plane** in Descartes' honor) is a familiar feature of middle school and high school algebra courses as students learn to graph linear, quadratic, exponential, and other functions. You may recall that it features two perpendicular axes, the horizontal *x*-axis and the vertical *y*-axis, which intersect at a point called the origin. We then label each point on the plane with an ordered pair of coordinates (*x*, *y*), where the *x*-coordinate tells us how far the point is from the origin (0, 0) in the horizontal direction and the *y*-coordinate gives the distance from the origin in the vertical direction. For example, the point (-3,-1) is located 3 units to the left and 1 unit down from the origin.

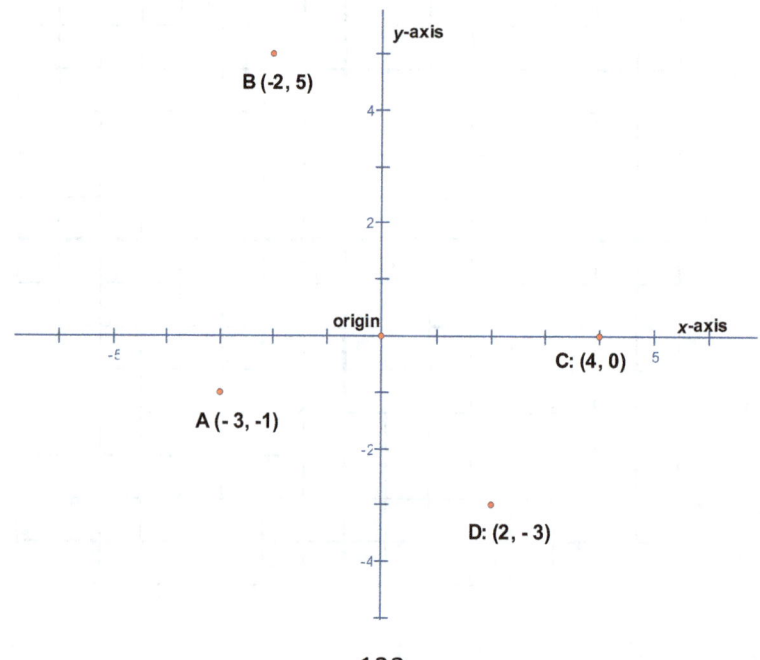

Using the Pythagorean Theorem we can find the distance between any two points on the Cartesian plane. For example, let's find the distance between points A and D in the picture above. The line segment AD is the hypotenuse of a right triangle with a horizontal leg of length 5 (2 – (-3)) and a vertical leg of length 2 ((-1) – (-3)). So the square of the distance between A and D is $5^2 + 2^2$ = 25 + 4 = 29 and the distance between A and D is $\sqrt{29}$. *Find the distance between (2, -3) and (4, 0).*

There are several facts about lines on the Cartesian plane that are useful to recall. One is that every line has a slope, which is a measure of its inclination with the x-axis. The idea of slope is that it is the amount you need to move in the y-direction to stay on the line for a one unit change in the x-direction. So think about this. *What does a slope of 7 mean? Sketch a line with that slope.*

What does a slope of -¼ mean? Sketch a line with that slope.

We can calculate the **slope** (m) of a line by using the coordinates of two points that lie on the line with the formula

$$m = \frac{y_2 - y_1}{x_2 - x_1}$$ where (x_1, y_1) and (x_2, y_2) are the coordinates of the two points.

Just to jog your memory, compute the slope of the line containing the points (4, 0) and (-2, 5).

If two lines are **parallel**, then they will make the same angle with the x-axis (a transversal) and so will have the same slope – and vice versa, if two lines have the same slope, then they are parallel.

Think about how you could make an argument for this fact. This turns out to be a very useful observation. If we need to show that two lines are parallel, we can simply calculate their slopes and show that they are equal. (Remember this when you get to the homework problems.)

What if two lines are perpendicular? How are their slopes related? It turns out that the slopes of perpendicular lines also have a numerical relationship. The product of the slopes of **perpendicular** lines is always -1. *Think about how you could make an argument for this fact. What would be the slope of the perpendicular to the line containing the points (4, 0) and (-2, 5)?*

And here is the last useful "fact" about using coordinates on the Cartesian plane that we need for our work.

The coordinates of the **midpoint of the line segment** connecting (x_1, y_1) and (x_2, y_2) are

$$\left(\frac{x_1 + x_2}{2}, \frac{y_1 + y_2}{2}\right).$$

Make an argument for this fact. What are the coordinates of the midpoint of the line segment connecting the points (4, 0) and (-2, 5)?

So what does all of this have to do with using algebra to solve geometric problems? That was the genius of Descartes' invention. We'll show you an example.

Consider the following geometric problem: Show that the segments joining the midpoints of the opposite sides of a quadrilateral bisect each other.

So we have any quadrilateral, nothing special about it, but if we connect the midpoints of its opposite sides, those segments will cut each other into two equal length pieces. *Draw a sketch to see that this seems true.*

Now, let's see if we can prove this using the structure of the Cartesian plane to help us out.

Our first step is to choose four random points and let them be the vertices of our quadrilateral – remember nothing special allowed – no parallel sides, no congruent sides, etc. But we *can* choose some easy-to-use points (such as the origin) for two of our points. (The problem-solver can lay down the structure wherever we like.)

We will label our points with coordinates (0, 0), (a, 0), (b, c), and (d, e). Even though we have to place these points in particular spots on our diagram, we are making no assumption about the actual values of *a, b, c, d,* and *e*.

Next we'll connect our four points to make the quadrilateral and then calculate the coordinates of the midpoints of each of the sides using the midpoint coordinate formula we talked about earlier. *Carry out these calculations for yourself. Do you get the same results?* Our diagram now looks like this:

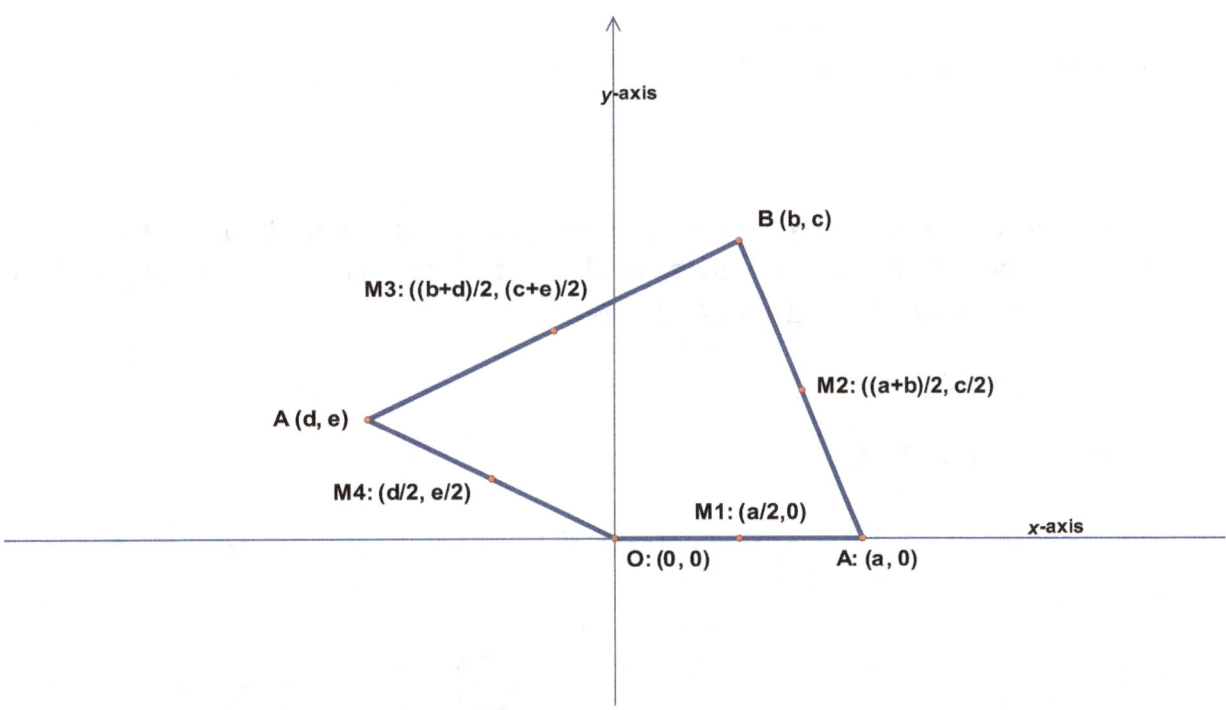

Lastly, we are interested in the two segments which join midpoints of opposite sides, that is segment M_1M_3 and segment M_2M_4. We need to show that they bisect each other at their point of intersection. *Think carefully for a few minutes – how can we show this? There are several approaches that will do the job, but some are easier than others. Decide on a method that makes sense to you and finish the proof before reading any further. (Hey, really do it.)*

One way to show that M_1M_3 and M_2M_4 bisect each other is to find the equation of each line and solve this system of two equations for the coordinates of their common point (let's call it *M*). Then we would find the distance from that point to each of the points M_1, M_2, M_3, and M_4. If the distance from *M* to M_1 and the distance from *M* to M_3 were equal and if the distance from *M* to M_2 and the distance from *M* to M_4 were equal, we are done. *Explain why showing that these pairs of distances are equal do show that M_1M_3 and M_2M_4 bisect each other.*

Another easier way might be to argue like this: If M_1M_3 and M_2M_4 bisect each other, then the midpoint of each segment must have the same coordinates. *Explain the logic of this statement before continuing.* So we can simply find the coordinates of the midpoint of each segment and demonstrate that these two midpoints are indeed the same point. We like this approach because it is simpler to carry out. So here are our calculations. *Make certain you can get the same results.*

Coordinates of the midpoint of M_1M_3 = $\left(\dfrac{\dfrac{a}{2}+\dfrac{b+d}{2}}{2}, \dfrac{0+\dfrac{c+e}{2}}{2}\right) = \left(\dfrac{a+b+d}{4}, \dfrac{c+e}{4}\right).$

Coordinates of the midpoint of M_2M_4 = $\left(\dfrac{\dfrac{a+b}{2}+\dfrac{d}{2}}{2}, \dfrac{\dfrac{c}{2}+\dfrac{e}{2}}{2}\right) = \left(\dfrac{a+b+d}{4}, \dfrac{c+e}{4}\right).$

So here is the point: we just took a purely geometric problem, translated it to an algebraic problem (or, said another way, we imposed an algebraic structure) and then we used algebra to solve it. We call this approach **analytic geometry**.

Connections to Teaching:

> *In grades 6-8 all students should use coordinate geometry to represent and examine the properties of geometric shapes.*
>
> National Council of Teachers of Mathematics
> Principles and Standards for School Mathematics, p. 232

Representational systems are structures that help us to make sense of geometric objects. Examples include grids, the coordinate plane, lines of longitude and latitude on a globe, maps and contour maps. Of course, the most important of these in middle grades mathematics is the coordinate plane because it lays the groundwork for graphing algebraic relationships. Here is the relevant Common Core State Standard for Geometry for students in grade six:

> 6.G.3 Draw polygons in the coordinate plane given coordinates for the vertices; use coordinates to find the length of a side joining points with the same first coordinate or the same second coordinate. Apply these techniques in the context of solving real-world and mathematical problems.

Copyright 2010. National Governors Association Center for Best Practices and Council of Chief State School Officers. All rights reserved.

Make up an example of a real-world problem that would have your students apply the techniques described above.

You will teach some analytic geometry. You will teach your middle grades students that the geometric object called a line can be described by an equation. It has the form $y = mx + b$ where m tells how steep the line is and b give its position on the coordinate axes (b is called the y-intercept).

You will also teach your students that a parabola (also a geometric object - just wait until the next *Class Activity*) has an equation of the form $y = a(x - k)^2 + h$, where (k, h) gives the vertex of the parabola and a tells how "fat" the parabola is. *Fill in the values of a, k and h (just make them up) and then graph the equation on your calculator. Now change a value and do it again.*

Homework:

Each problem that I solved became a rule which served afterwards to solve other problems.

Rene Descartes

1) Do all the italicized things in the *Read and Study* section.

2) Do the *Connections* problems.

3) If you haven't already done so, prove that two lines are parallel if and only if they have the same slope.

4) Apply the Pythagorean Theorem to the points (x_1, y_1) and (x_2, y_2) to derive the formula for finding the distance between two points on a coordinate grid:

$$d = \sqrt{(x_2 - x_1)^2 + (y_2 - y_1)^2}$$

5) Use the methods of analytic geometry to show that the four midpoints of any quadrilateral always form a parallelogram.

6) Use analytic geometry to determine the curve that the midpoint of a ladder makes as the top of the ladder slips down a wall and the bottom of the ladder moves away from the wall. (*Hint*: Draw a diagram. Would this be the same as finding the set of all midpoints of segments of ladder length whose endpoints are on the x- and y-axes? You can have the ladder be of length 1 to simplify your calculations.)

7) Use the methods of analytic geometry to show that the diagonals of a rectangle are congruent.

8) Use the methods of analytic geometry to show that the diagonals of a rhombus are perpendicular.

9) If you haven't already done so, use the methods of analytic geometry to find the solution to the Buried Treasure problem from the *Class Activity*.

Class Activity 19: Plague of Locus

There are no sects in geometry.

Voltaire

1) Imagine two infinite (hollow) cones with their tips touching at one point. Now think of all the ways you could slice through those cones with a plane. What are the possible curves (or other objects) that could result (don't look on the back of this sheet until you've done this). Sketch a picture of each.

2) Each of these objects has a geometric definition (that we call the **locus definition**), and if you apply analytic geometry to that definition you get the familiar algebraic formula for the object. Here's an example: You probably decided that a circle would result if you sliced through just one of the cones with your plane parallel to the "base". The locus definition of a circle is this:

A **circle** is the set of points in the plane that are equidistant from a given point in the plane (called the locus).

Now, if we put down a Cartesian coordinate system on that plane and call the center of our circle (h, k) and the radius r, we can find an *equation* that must be satisfied by all the points (x, y) that lie on that circle. Draw a sketch and then derive that equation.

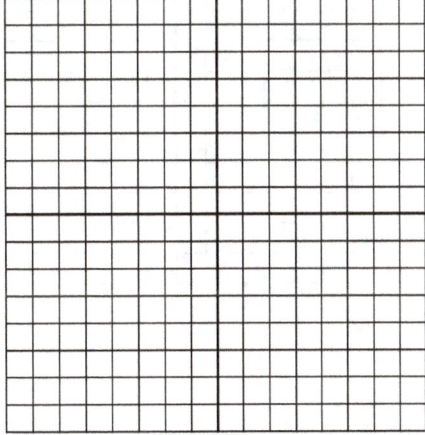

(This activity is continued on the next page.)

3) The locus definition for a parabola is this: A **parabola** is the set of all points in the plane that are equidistant from a given point (the focus) and a given line (called the directrix).

Use the definition above to sketch a parabola with focus (3, 6) and directrix, $y = 2$ on the graph paper below. Now find the equation for that parabola.

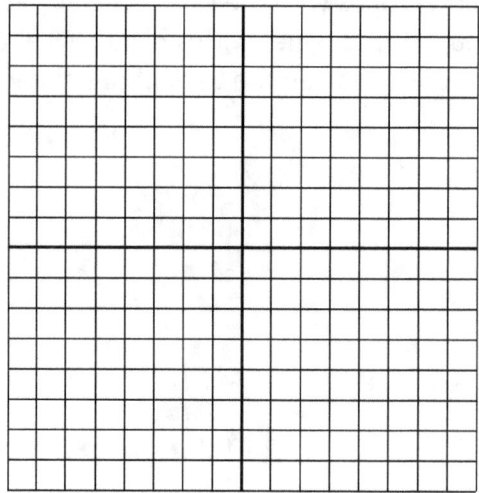

4) The locus definition for an **ellipse** is the set of all points in the plane such that the *sum* of the distances from two given points (the foci – that's plural for focus) is constant.

Use the definition to sketch a picture of an ellipse with foci (-2, 0) and (2, 0) and a constant sum of 7 on your graph paper. You do not need to find its equation.

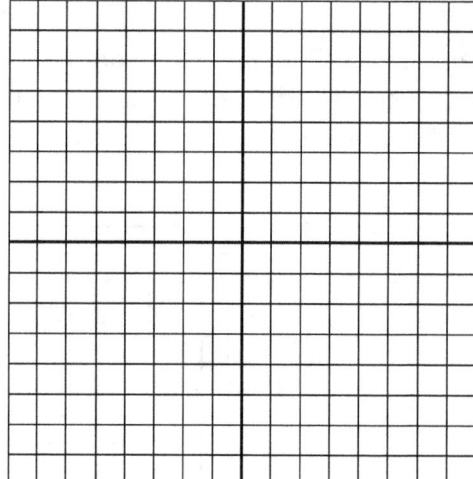

(This activity is continued on the next page.)

5) The locus definition of a **hyperbola** is the set of all points in the plane such that the *difference* of the distances from a point on the hyperbola and two given foci is constant.

   Use the definition to sketch a picture of an hyperbola with foci (0, 0) and (6, 0) and a constant difference of 4 on your graph paper. You do not need to find its equation.

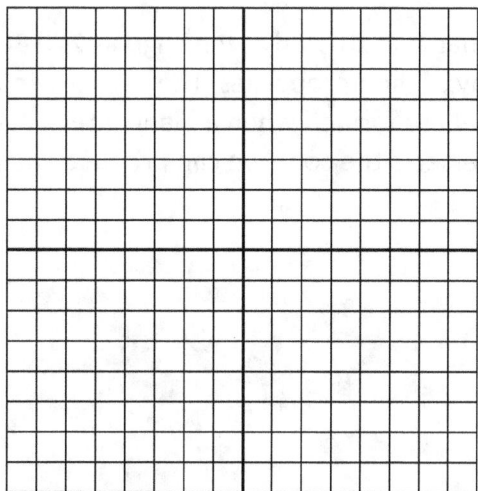

6) You may have decided that a point and a line could also be formed by slicing your infinite cones in problem #1. We can think of those objects as "degenerate forms" of these four objects we've already listed. For example, a point is a degenerate circle (the circle with zero radius). What is a line? Explain.

Read and Study:

Inspiration is needed in geometry, just as much as in poetry.
                                            Aleksandr Sergeyevich Pushkin

The **conic sections** were named and studied as long ago as 200 BC, when Apollonius of Perga undertook a systematic study of their properties. They are the four curves (the circle, ellipse, hyperbola, and parabola) that are formed when a plane intersects a double cone. By varying the angle at which the plane intersects the cone we can produce each of them, as shown below.

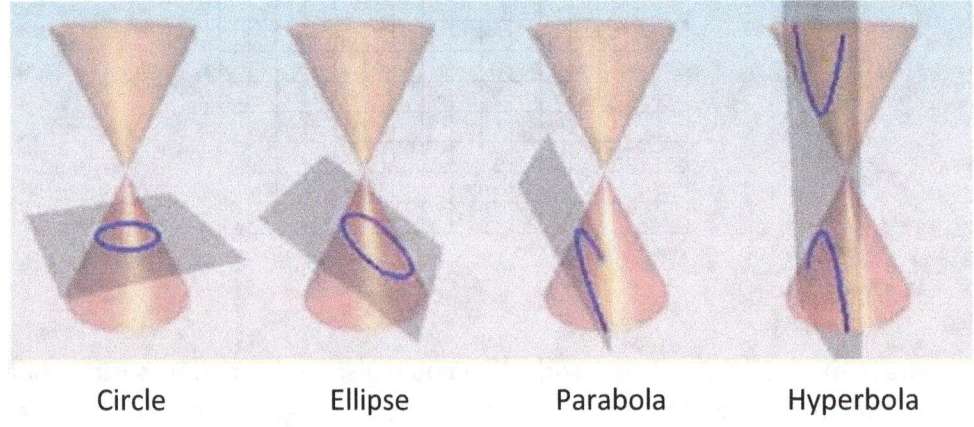

Circle   Ellipse   Parabola   Hyperbola

(Illustrations taken from http://math2.org/math/algebra/conics.htm.)

In the Class Activity you found that of these curves can be defined using a **locus** definition (a definition that describes the curve as a set of points in the plane). For example, you were asked to use the Euclidean distance formula and this definition of a circle to determine the general equation of the circle of radius r and center (h, k). Now we will further discuss (with illustrations) the ellipse, the hyperbola, and the parabola:

Given two points F_1 and F_2, an ellipse is the set of points P in the plane such that the sum of the distances from P to F_1 and F_2 is constant. This means that if we take any point P on the ellipse and measure the distance between P and F_1 and the distance between P and F_2, then when we add these two distances together we will always get the same sum. *What would happen to the shape of this ellipse if we moved F_1 and F_2 closer together but kept the given distance constant?*

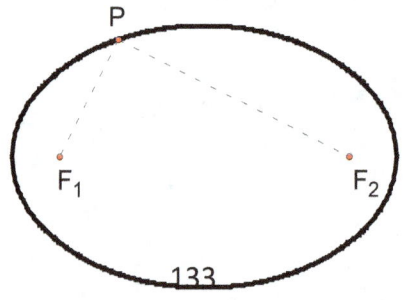

A **hyperbola** is the set of points P in the plane such that the difference of the distances from P to F_1 and F_2 is constant. *What would happen to the shape of the hyperbola if we moved F_1 and F_2 closer together but kept the given difference constant?*

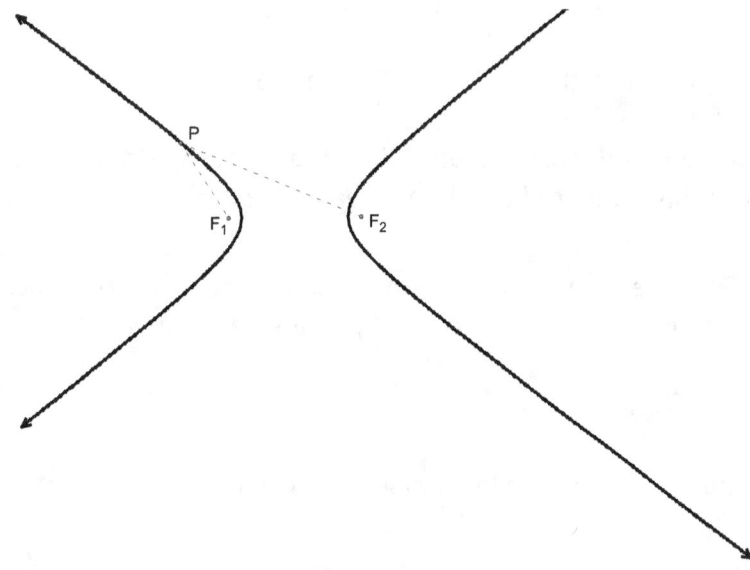

A **parabola** is the set of points P in the plane such that the distance from P to a given point F is equal to the distance from P to a given line m. (Recall that Point F is called the **focus** of the parabola and line m is the **directrix.**) *What would happen to the shape of the parabola if we moved the directrix further from the focus? What would happen to the parabola if we changed the directrix to a vertical line?*

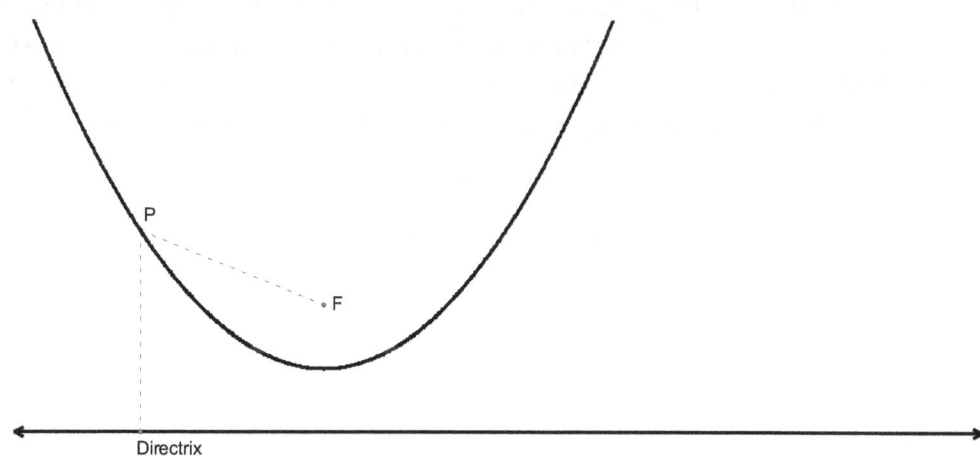

Homework:

All truths are easy to understand once they are discovered; the point is to discover them.
Galileo Galilei

1) Do all of the italicized things in the *Read and Study* section.

2) The Common Core State Standards lists the following standard for students in grade seven. How does this standard fit with the ideas described in this section?

> 7.G.3 Describe the two-dimensional figures that result from slicing three-dimensional figures, as in plane sections of right rectangular prisms and right rectangular pyramids.

Copyright 2010. National Governors Association Center for Best Practices and Council of Chief State School Officers. All rights reserved.

3) Explain how to form a line by intersecting a plane with a pair of infinite cones.

4) In analytic geometry a **line** is the set of all points (x, y) that satisfy the equation $ax + by + c = 0$, where both a and b are not zero. Find the slope and y-intercept of the line in terms of the real number parameters a, b and c. What happens when $a = 0$? When $b = 0$? When $c = 0$?

5) Find the equation of a circle with center (2, -4) and radius 5.

6) Suppose the equation of the directrix of a parabola is $y = -3$ and the point $F = (0, 5)$ is its focus. Find its equation.

7) An ellipse can be modeled using two stick pins (one at each focus) and a length of string (equal to the sum of distances from the ellipse to the foci). Experiment with this method to create various ellipses. What happens when the length of string stays the same but you vary the position of the foci? What happens when you keep the foci fixed but vary the length of the string? Is there a minimum length of string necessary?

Summary of Big Ideas from Chapter Two

If an idea's worth having once, it's worth having twice.
Tom Stoppard

- The van Hiele levels describe a progression of geometric understanding.

- It is important for your students to make sense of the formulas for area and volume.

- There are three rigid motions of the plane: rotation, reflection, and translation.

- If glide reflection is considered its own motion, then the composition of any two rigid motions is another rigid motion.

- Creating tilings and tessellations is an activity that can be adapted to every grade level – very young children can create patchwork quilts from construction paper using only rectangles or squares or isosceles right triangles. Older students can make more complex artwork using the mathematical concepts of congruency and transformations.

- Two figures are similar if there is a sequence of rigid motions and a dilation of the plane which maps one figure onto the other.

- Analytic geometry involves taking a geometric problem and translating it into an algebraic problem. It is a very useful proof technique.

- We can use analytic geometry to help us describe figures like parabolas and hyperbolas.

Chapter Three

Exploring Strange New Worlds: Non-Euclidean Geometries

Class Activity 20: Life on a One-Sided World

Only those who attempt the absurd will achieve the impossible. I think it's in my basement... let me go upstairs and check.

M. C. Escher

Cut several 1-inch wide strips (the long way) from a blank sheet of 8½ by 11 inch paper. With one strip, tape the one-inch ends together to form a cylinder. With another, make a half-twist and then tape the one-inch ends together to form a two-dimensional surface called a Möbius strip. Save the remaining strips for additional examples, as needed.

1) How many "sides" does the cylinder have? The Möbius strip? What makes the difference? How many sides does a strip made with 2 half-twists have? One with 3 half-twists? How many sides does a strip with 46 half twists have? One with 511?

2) How many edges does the cylinder have? The Möbius strip? How many edges do the strips with 2, 3, 46, or 511 half twists have? Explain the difference.

3) Is there a connection between the number of sides and the number of edges? What about between the number of half-twists and the number of sides? Between the number of half-twists and the number of edges? Explain all of this.

4) What happens when you cut a cylinder down the middle? What happens when you cut a Möbius strip down the middle? Think about it before you do it! Then explain precisely. (e.g., What pieces result? How are they linked? How are the sizes related?) What if you cut these pieces down the middle (i.e., cut the original strip into fourths)?

Read and Study:

...by natural selection our mind has adapted itself to the conditions of the external world. It has adopted the geometry most advantageous to the species or, in other words, the most convenient. Geometry is not true, it is advantageous.

Henri Jules Poincare

The **Möbius strip** is named after August Ferdinand Möbius, a nineteenth century German mathematician and astronomer, who was a pioneer in the field of topology. (By the way, topology is the study of spaces where the questions of interest are things like: *Does the space have holes in it? Is it connected?* Sometimes topology is called rubber-sheet-geometry because in topology one space is considered the same as another space if it can be bent or stretched into the other space. For example, in topology, a cube and a sphere are the same space but a donut and a sphere are not. *Why not?*) Möbius, along with his contemporaries Bolyai, Lobachevsky, and Riemann, turned the world of Euclidean geometry upside down, inside out, and every which way but flat.

The Möbius strip is a simple surface with surprising properties. A true Möbius strip is a two-dimensional surface (as if our strip of paper had no thickness whatsoever) with only one side and only one boundary edge. If we restrict ourselves to a small section of the Möbius strip, the geometry there is the same as it is on the flat (Euclidean) strip of paper from which it was formed. However, when we consider the entire Möbius strip, the geometry is quite different. Not only is it a surface with only one side and one edge, but it is also what we call non-orientable. If an amoeba living on the surface made a trip around the entire Möbius strip, it would return to its starting point as a mirror image of itself! *Think about this*. This kind of thing doesn't happen on a Euclidean surface such as the cylinder.

These surprising properties make the Möbius strip quite useful in the "real world." Giant Möbius Strips have been used as conveyor belts (to make them last longer, since "each side" gets the same amount of wear) and as continuous-loop recording tapes (to double the playing time). In the 1960's Sandia Laboratories used Möbius Strips in the design of versatile electronic resistors. Free-style skiers have christened one of their acrobatic stunts the Möbius Flip. The international symbol for recycling is a Möbius strip.

We get even more surprising results if we glue the two edges of a cylinder or a Möbius strip together. Try to imagine bringing the two open ends of the cylinder towards each other (it helps if you are imagining a long skinny cylinder – like a paper towel tube). *What shape would result?*

Mathematicians call this shape a **torus**. Bagels and inner tubes are two examples. Now imagine gluing two Möbius strips together edge to edge. You have to just imagine it – it is physically impossible to accomplish the gluing in three-dimensional space without tearing the Möbius strips. he Klein Bottle

What results is called the **Klein Bottle** – a surface whose inside is its outside! The apparent self-intersection you see in the following picture is misleading – the Klein bottle exists in 4-dimensional space with no self intersections. It was first described in 1882 by the German mathematician Felix Klein.

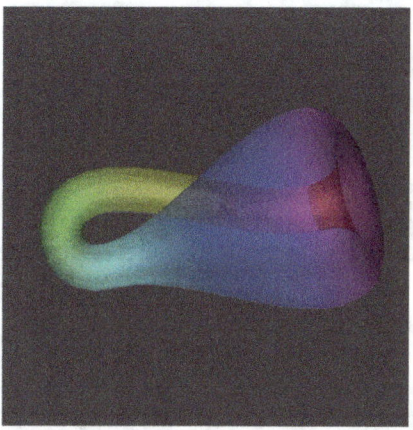

Illustration from http://www.geom.uiuc.edu/zoo/toptype/klein/standard/gifs/trans.gif .

Cylinders, Möbius strips, tori (plural of torus), and Klein bottles can all be represented by a "flat" rectangle with appropriate gluing instructions for the opposite edges. The opposite edges with arrows are to be glued together with arrows matching. We call these **identification spaces** for the objects.

This is the same idea used in video games where the space craft or robot or whatever leaves the screen on the left side and returns on the right or leaves on the top and returns from the bottom and vice versa. *Study the gluing directions for each object and explain how they match the physical models you have made (or imagined, in the case of the Klein bottle).*

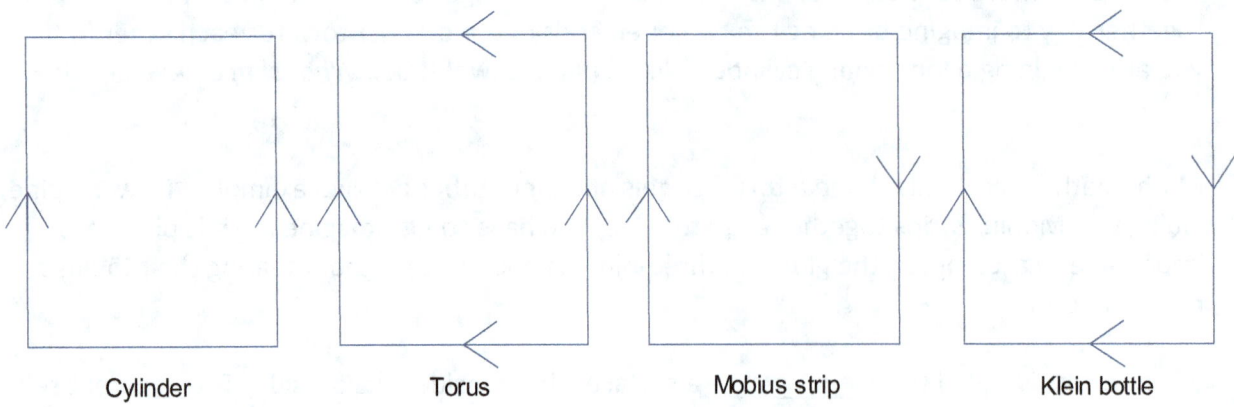

Connections to Teaching:

If you would thoroughly know anything, teach it to others.

Tyron Edwards

You middle grades students will like to explore the geometry of the cylinder, torus (donut), Möbius strip and Klein bottle (among others) through activities, puzzles and games. For example, here is a word search on a torus. See if you can find all of these words: possum, panda, jaguar, camel and llama.

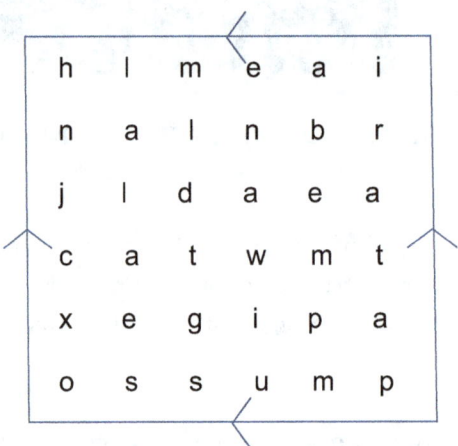

Homework:

Whenever you are asked if you can do a job, tell 'em, 'Certainly I can!' Then get busy and find out how to do it.

Theodore Roosevelt

1) Do all the italicized things in the *Read and Study* section.

2) Do the word search from the *Connections* section. Then see if you can make up a word search (the same size as the one above with at least five words to find) on a Klein Bottle.

3) Use an identification space to predict what would happen if you cut a Möbius strip into thirds. Then, check it out. If your predictions were wrong, try to figure out where you made an error in thinking. Why do the actual results make sense?

4) Predict what would happen if you cut a strip with three half-twists in half down the middle. Check it out. If your prediction was wrong, try to figure out where you made an error in thinking. Why do the actual results make sense? The resulting object is known as a trefoil knot. (Knot theory is another fun area of mathematics related to geometric ideas.)

5) Use the flat models of the cylinder, the Möbius strip, the torus, and the Klein bottle to create tic-tac-toe game boards. Play several games on each surface. Don't forget to include the gluing instructions in your strategy. How does the game change on each surface? What strategies can you use to win in each case? Is there a surface on which you can guarantee a win by going first? By going second? Is there a surface on which the game always results in a tie? (Assume two competent players and that neither makes a mistake.)

6) Three amoebas, Apox, Brillo, and Cheesy, line up for a race on a virtual Möbius strip swimming pool. All three swim up the middle of their lanes at exactly the same speed. Which amoeba will return to his or her own starting point first? Why?

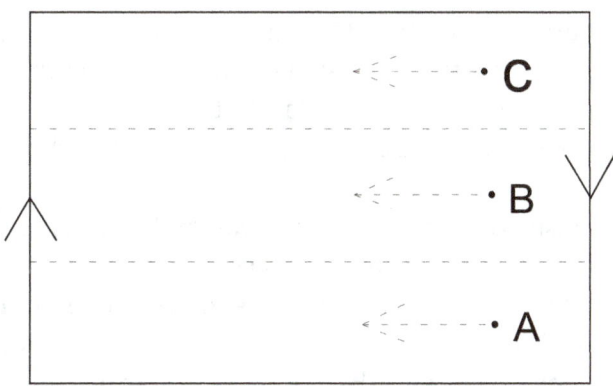

Class Activity 21: Life in a Taxicab World

> *To fully appreciate Euclidean geometry, one needs to have some contact with a non-Euclidean geometry.*
>
> Eugene F. Krause, Taxicab Geometry

Terrance and Sasha live in Perfection City where all streets intersect at right angles and are evenly spaced. A model of Perfection City is the Cartesian plane with streets represented by vertical lines at all integer values of the *x*-axis and avenues represented by horizontal lines at all integer values of the *y*-axis. Unlike the Cartesian plane, Perfection City is not infinite in size; we will focus on the heart of the city contained within the grid $-10 \leq x \leq 10$ and $-10 \leq y \leq 10$. Terrance works at the public library located at the corner of 3rd Street East and 1st Avenue South and Sasha teaches math at Perfection High School located at the corner of 5th Street West and 9th Avenue North. (Notice that only the even numbered streets and avenues are shown on this grid.) Locate the library and the High School on the grid.

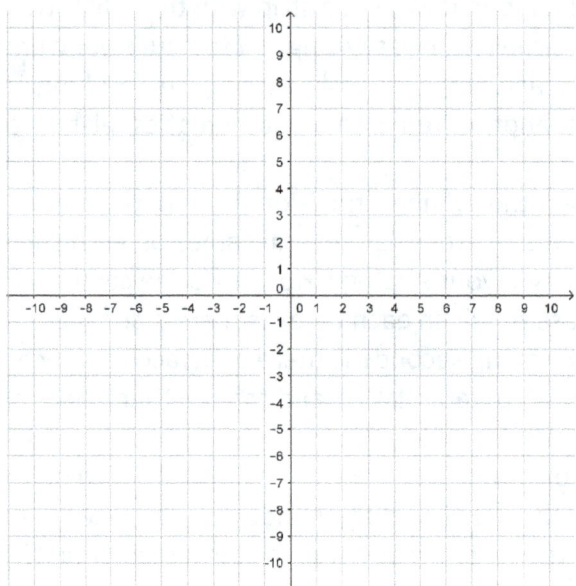

1) How far apart are the library and the high school (as the crow flies)? Staying on the streets, how far must either of them walk to meet the other at their workplace? Sasha likes to walk a different route each day, but she also wants to walk the shortest distance possible. For how many days can she make the walk without repeating a route?

2) Terrance and Sasha decide to meet halfway for lunch. Where is this halfway point? Is there more than one halfway point? Mark all of the halfway points on the grid with the letter *M*. Now suppose that none of these intersections contain an eating place satisfactory to both of them, where else could they meet for lunch so that each of them has the same length walk? Mark all of these points on the grid with the letter *P*. What is the mathematical description of the "line" which joins all of the points labeled *M* or *P*?

Read and Study:

Geometry, which is the only science that it hath pleased God hitherto to bestow on mankind.

Thomas Hobbes

Suppose we take the Euclidean plane and change nothing except our definition of *distance*. Points are still points; lines are still lines; and angles are still measured in the familiar way. But we will no longer use the "as the crow flies" definition of distance based on the Pythagorean Theorem. Instead we will measure the distance between two points by finding the sum of the vertical distance and the horizontal distance between the two points. In other words, we will measure distance in the same way that we measured the length of Terrance and Sasha's walks in the class activity. We can use the following formula to determine this new distance d_T between the two points (x, y) and (u, v):

$$d_\tau = |x - u| + |y - v|$$

A geometry with this new way of measuring distance is often called **taxicab geometry** because this formula gives the distance a taxi goes if it travels only along north-south and east-west streets, as in Perfection City.

Why would we suddenly want to change the definition of distance? After all, Euclidean geometry has served us well for the last 2000 years. There are a few possible answers to this question. The most obvious one is suggested by the name of taxicab geometry. Euclidean geometry measures distance "as the crow flies," but this doesn't always provide a good model for a real-life situation, particularly in cities, where one is only concerned with the distance their car will need to travel.

Another reason for studying taxicab geometry is that it is a simple non-Euclidean geometry. Taxicab geometry is fairly intuitive and requires less mathematical background than other geometries; in short, it is a good example of a non-Euclidean geometry for middle school students.

Let's examine this new definition of distance more closely. *Really do these things in italics below.*

First, calculate the normal Euclidean distance between points (2, 5) and (4, 1) and then find the taxicab distance between these two points. What about the points (2, 5) and (2, 1)? The points (2, 5) and (4, 5)?

Okay, what did you find? Are there pairs of points for which the "normal" distance and the taxicab distance between them are equal? If so, generalize the relationship between pairs of points for which this is true. When the taxicab distance and the Euclidean distance are not equal, which one is greater? Will this always be the case? Why?

All of Euclid's postulates hold in taxicab geometry, but definitions based on distance can look different. For example, let's consider circles.

What would a circle of radius 5 centered at the origin look like in taxicab geometry? *Think about the definition of a circle and the taxicab definition of distance and sketch the taxicab circle of radius 5 on the following pair of axes.*

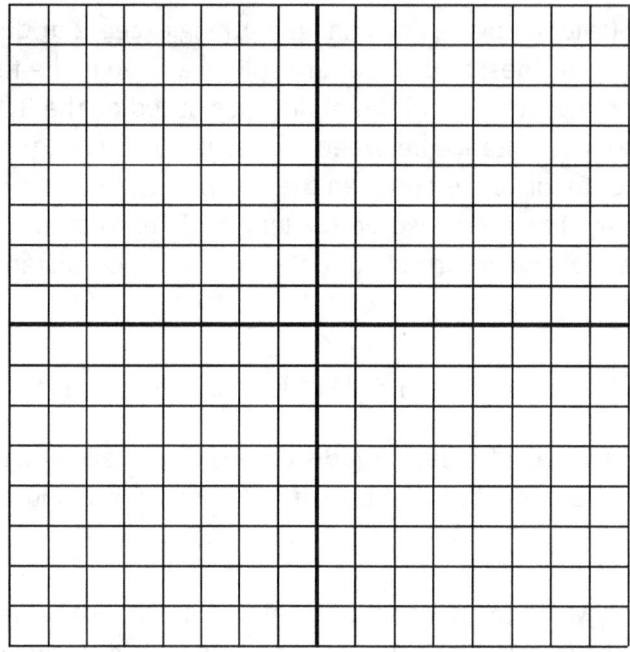

What is the circumference of this taxicab circle? (Remember to measure it too using taxicab distance.)

Now, recall that p is defined to be the ratio of the circumference of a circle to its diameter. In Euclidean geometry, p is an irrational number (approximately equal to 3.1416). *What is a reasonable value for the ratio "p" in taxicab circles? Why? Will it be a constant value for all taxicab circles? Explain.*

Many other familiar objects also "look" different in taxicab geometry. In the homework you will be asked to explore the shape of taxicab squares, equilateral triangles, and the conic sections.

Some of our familiar Euclidean results are no longer valid in Taxicab geometry. For example, consider the triangle congruence theorem, Side-Angle-Side (SAS). *Use the following two triangles and the formula for d_T to create a counterexample showing that SAS is not true in taxicab geometry.*

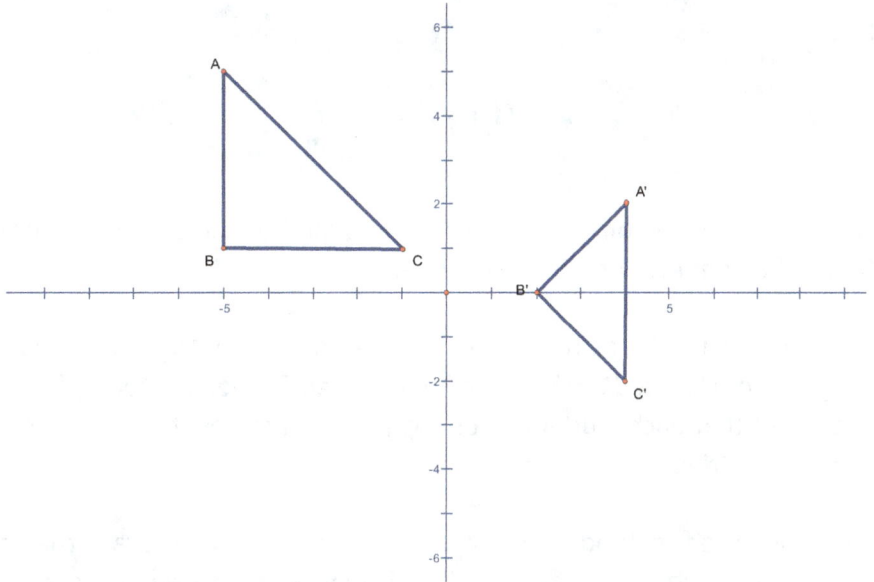

What does this example say about the Pythagorean Theorem in taxicab geometry? Does it hold? What about the other triangle congruence theorems? Will any of them be valid or can you find counterexamples for them as well? Check it out for some examples.

In Euclidean geometry, the set of all points equidistant from two given points is the perpendicular bisector of the line segment joining the two points. What will the "perpendicular bisector" of a line segment look like in taxicab geometry? In the second part of the class activity, all of the points labeled *M* and *P* were equidistant from the (L)ibrary and the high school (*HS*). So the line segments joining these points form the "perpendicular bisector" of the line segment joining *L* and *HS*.

Now consider the taxicab perpendicular bisector of the segment joining (2, 2) and (-1, -1). How does this "perpendicular bisector" differ from the Euclidean one? In what ways is it similar? When will a taxicab "perpendicular bisector" look like a Euclidean perpendicular bisector? When will it be different?

Connections to Teaching:

Whoever ceases to be a student has never been a student.
Georg Iles

Typically study of non-Euclidean geometries is not explicitly part of the upper elementary or middle grades curricula. However, there are pieces of these geometries that will help students to understand maps and map-making. There are many good problem ideas in Taxicab geometry for the middle-grades at the website http://emat6000taxicab.weebly.com/teacher-resources.html.

Homework:

Say not, 'I have found the truth,' but rather, 'I have found a truth.'
Kahlil Gibran

1) If you haven't already done so, go back and do all the italicized things in the *Read and Study* and the *Connections* sections above.

2) Perfection City actually has three high schools: Perfection High School located at (-5, 9), Ideal High School located at (8, -1) and Idyllic High School located at (0, -7). Draw the school boundaries so that each student attends the school closest to his or her home, as the taxi (or school bus) drives.

3) *Model Burger*, the fast-food chain, wants to open a new restaurant that is centrally located so that it is the same taxicab distance from each of the three high schools. Where should it be located?

4) Terrance and Sasha need to find an apartment so that that the sum of the distances that the two of them will walk to work should be no more than twenty-four blocks. Draw the boundary of their search area. Which of the conic sections are they using to define the search area?

5) When Terrance and Sasha were unable to find an apartment, they next agreed that neither of them should have to walk more than four blocks farther than the other in order to get to work. Now where can they look? Which of the conic sections are they using to define the search area this time?

6) In the reading, you discovered that taxicab circles look like Euclidean squares. What do taxicab squares look like? Use the definition of a square and the taxicab definition of distance and draw three taxicab squares with a side length of four such that the figures are not congruent as Euclidean figures. What Euclidean shape do the taxicab squares have? Why does this happen?

7) Now experiment with taxicab triangles. Can you draw a regular triangle in taxicab geometry? Why or why not? How about a right triangle with sides of equal length? How about an isosceles triangle whose base angles are not congruent?

Class Activity 22: Life on a Spherical World

You can't comb the hair on a ball!

Mary Ellen Rudin

In this activity you will explore geometry on the surface of a Euclidean sphere by working with a physical model of a sphere (a ball) and a physical model of a line (a piece of string). You may need markers to draw lines on the sphere (or rubber bands to model lines) and a regular protractor to measure angles. Assume that the radius of your sphere is one unit.

1. Talk with your group and decide how you can use a piece of string to make a straight line – first on a flat sheet of paper (Euclidean model) and then on the sphere. *Take this seriously – it is important to have a valid model of a straight line before proceeding.* Relate what you have decided about straight lines on the sphere to the "lines" of longitude and latitude markings on a globe.

2. Draw a straight line on your sphere (not a segment but a line). How long is it? Now find a different straight line that is parallel to it. (Recall that lines are parallel if they have no points in common.) How many lines parallel to your original line can you find? Is the geometry of the surface of a sphere Euclidean? Why or why not?

3. Mark two points anywhere on the sphere. Draw the line segment (using the string method) between these two points. What do you notice? How many line segments can you find? Does it matter where the two points are in relation to each other? Experiment with various pairs of points and form a conjecture about line segments on a sphere.

(This activity is continued on the next page.)

4. Draw a small triangle and a large triangle (one that covers at least 1/8 of the surface area on your sphere). Make certain that the sides of your triangles are actually straight line segments by using the string method to construct the triangle. Determine a method to measure the angles of the triangles using your protractor and then measure each of the angles in both triangles. What is the angle sum of the small triangle? The large triangle? Now draw a medium-sized triangle and a really big triangle and measure their angle sums. Make a conjecture about the angle sum of a spherical triangle.

5. Draw a right triangle on your sphere. How can you make certain that you have a right angle? How many right angles can you have in one triangle? Can you a draw a triangle that has two right angles? Can you draw a triangle that has three right angles? Do you think the Pythagorean Theorem holds on a sphere? Why or why not?

6. Draw a line on the sphere and choose a point that is not on that line. How many perpendicular lines to your original line can you draw through that point? Are there points you can choose where there would be many perpendicular lines through that point? If so, describe these points and explain why you have more than one perpendicular to the line through those points.

Read and Study:

Mathematics, rightly viewed, possesses not only truth, but supreme beauty – a beauty cold and austere, like that of a sculpture

Bertrand Russell

So far you have studied two infinite non-Euclidean geometries, each created by one simple change to the familiar Euclidean geometry of the flat plane. In the Möbius strip we introduced a half-twist before gluing together one pair of opposite sides of a flat rectangle. In taxicab geometry we changed the definition of distance on the flat plane. Now we'll consider what happens when we introduce a constant positive curvature to the flat plane. The fact that the curvature is *positive* causes the plane to close up into a ball – the fact that the curvature is *constant* means that our ball is perfectly round (like a basketball and not a football). In fact, the Cartesian plane with constant positive curvature becomes the surface of a sphere – and this single change again affects the geometry in drastic ways.

For starters, we no longer have an infinite plane. The surface area of a sphere is finite and depends on the radius. Remember the formula, $A = 4\pi r^2$, for surface area of a sphere with radius *r*? This area can be quite small, as on a beach ball, or it can be quite large, as on the planet Jupiter, but it is always finite. This effectively means that the size of every geometric object drawn on the surface of a sphere has a limiting size – there is a largest circle, there is a longest line segment, and there is a biggest triangle.

Then there is the story about lines. In order to maintain the concept of straightness on the sphere, we have to use the fact that on a flat plane a straight line is the shortest distance between two points (represented by pulling a string tight between those two points). When you pulled the string tight against the surface of the sphere and went all the way around the sphere back to your starting point, you created a model of a straight line on the sphere.

This "straight line" is a **great circle**, a circle formed on the surface of the sphere by the intersection of a plane that goes through the center of the sphere. You will know that a circle on the sphere is a great circle if it cuts the sphere into two halves of equal area (two hemispheres). The equator on a globe is an example of a great circle. So are the lines of longitude - but not the latitude markings. On a sphere, great circles are lines; all other circles are just circles. So lines are also finite in length – in fact, all lines have the same length. *What is the formula for the length of a line on a sphere with radius r?*

Another surprising finding about lines on a sphere is that there are no parallel lines. All lines intersect, and in fact they all intersect in exactly *two* antipodal points. (**Antipodal points** are points that are at opposite ends of a diameter of the sphere, like the north and south poles.) Thus spherical geometry is non-Euclidean in a most basic way – *it does not satisfy Euclid's 5th Postulate about parallel lines*. Since there are no parallel lines, there can be no parallelograms, rhombi, rectangles, or squares either. A little bit later we will explore another argument for the fact that rectangles and squares do not exist in spherical geometry.

In the activity, you found that there are always two line segments between any two points on the sphere – and when those points are antipodal, there are an infinite number of line segments between them. Again, this is not at all like what happens in the flat plane. If the points are not antipodal, then one of the segments is shorter than the other and together the two segments compose the entire line between the two points. (The shorter one is called the **minor segment**. The longer one is the **major segment**.) If the points are antipodal, then every segment between them is equal in length to half the circumference of the sphere.

Since we can form line segments between points, we do have triangles on the sphere. But if we start with three points, there is more than one triangle we can form with those three points as vertices. So two triangles may share the same vertices, but have different length sides, different angle measures, and different areas. And in fact, even if we specify that the sides are to be the minor segments between the points, we still have two triangles of different area formed by those segments. *Did you see this when you were forming your triangles in the class activity? Stop now and use a ball to visualize exactly what we are saying.*

This is a good time to point out again the importance of carefully worded definitions. On the flat plane it is sufficient to say that a triangle is three non-collinear points and the line segments joining those points. On the spherical plane we must refine our definition to say that a triangle is three non-collinear points and the minor line segments joining those points, taking the interior of the triangle to be the smaller of the two areas enclosed by those segments. *Is it necessary to include the requirement that the points be non-collinear in the spherical definition? Can we place three points on the same line and choose line segments between them to form a triangle? What would be the area of such a triangle?*

Since any line on the sphere is a great circle, we can define the angle between two lines as the angle formed by the intersection of the two planes that create the great circles that are those lines. Since those two planes can intersect in any angle between 0° and 180°, we have the same angle measures on the sphere. In particular, we have angles of 90° between lines on the sphere and so we have perpendicular lines and right triangles. In the class activity, you investigated right triangles, and in particular, whether or not it was possible to have two or even three right angles within one triangle. *What conclusions did you make? Can you describe a triangle on the sphere that has two right angles? That has three right angles?*

A triangle with three right angles would have an angle sum of 270° so the fact that triangles in Euclidean geometry have angles sums of 180° must come from the 5th postulate. Change your axioms, and you change your theorems. *What did you find to be the angle sums of the triangles you formed in the class activity? What was the smallest angle sum you found? The largest? What would be the largest angle sum possible? Why?* Of course, your measurements with a protractor were approximate, as are all measurements, but you should have found that your angle sums were *all* larger than 180° and that as the area of the triangle became larger, so did the angle sum. In fact, it is an amazing feature of spherical geometry that the angle sum of any triangle is greater than 180° and that the area of a triangle is equal to its angle sum (in radians) minus p. To try and understand this, consider a type of polygon that does not exist in Euclidean geometry, a two-sided polygon called a **biangle** or a **lune**.

Since every pair of lines on the sphere intersects in two points, we do have a polygon with two sides and two vertices (which will be antipodal). Why is it called a lune? The name comes from the Latin word *luna*, which means moon. Think about the part of the moon that is seen at any time. That portion has to be both in the hemisphere which is illuminated by the sun and in the hemisphere that is visible from the earth. The intersection of two hemispheres is precisely a lune. Every pair of lines will form two pairs of congruent lunes (similar to the two pairs of congruent vertical angles formed by intersecting lines on the flat plane). *Study the diagram below to make certain you understand this definition.* One of the four lunes formed is shaded with vertical hatching. *Do you see the lune congruent to it? See the other pair of congruent lunes? What will be the area of the shaded lune if the angle between the two sides is 30° (p/6 radians) and the radius of the sphere is one unit?*

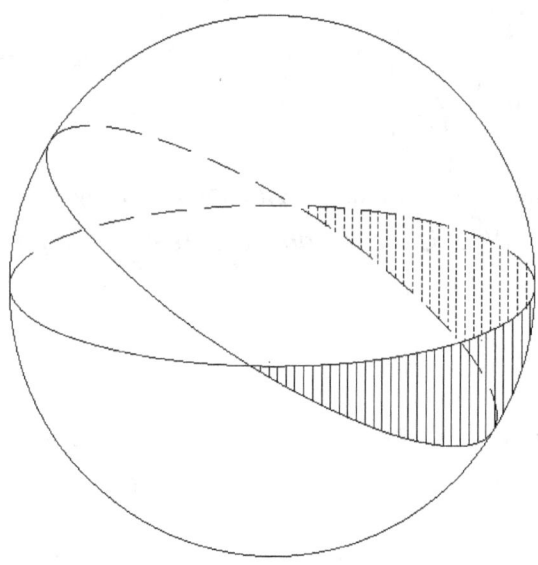

152

[Here you might need a quick reminder about radian angle measure. As we mentioned earlier, assigning 360 degrees to one full rotation is just arbitrary. Another standard way to measure angles is using the length of the circular arc that the angle sweeps out, using the radius as the unit of measure. In this way, 1 radian is the angle that sweeps out an arc length equal to the length of the radius, as illustrated below:

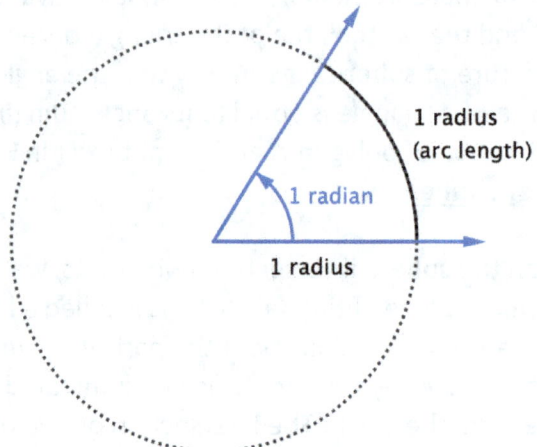

If the radius has a length of 1, then the arclength is the measure of the angle (in radians). So, in radians, the measure of the 80 degree angle shown below would simply be the length of the arc.

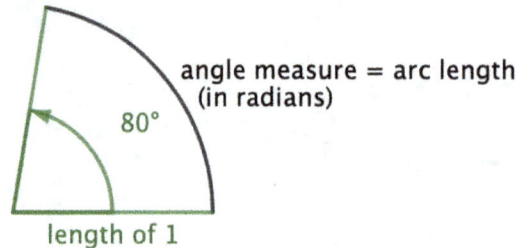

Now in a full rotation the length of the arc is **2π** radians. (*Why is that?*) So our 80 degree angle is a little less than $\frac{1}{2}$ π radians. *What is the radian measure of an angle that measures 45 degrees? 180 degrees?*]

Now we will get back to finding the area of a spherical triangle. It will help a lot if you have a ping pong ball or tennis ball or some other ball that you can write on to follow along (and to *think* along) with us.

Study the figure below until you are comfortable explaining how triangle ABC is formed by the intersection of lune AA', lune BB', and lune CC'. Notice that there is a mirror image triangle A'B'C' formed on the back side of the sphere. We will assume that the radius of the sphere is one unit and that ∠CAB = **a** radians, ∠ABC = **b** radians, and ∠BCA = **g** radians.

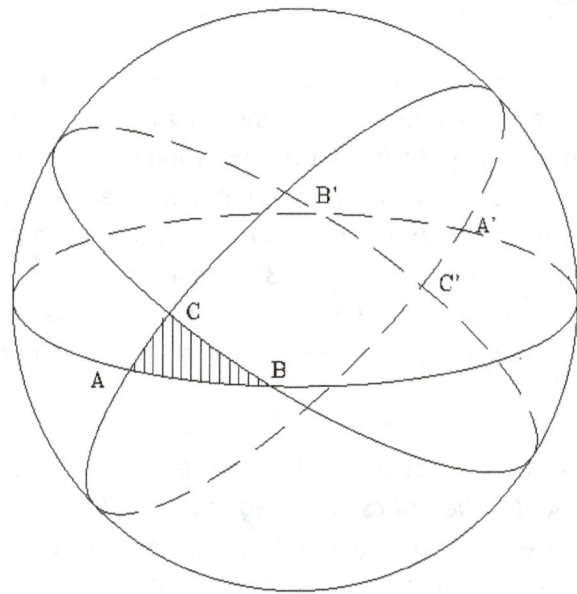

The area of the entire sphere is 4π. The area of each lune is equal to twice its angle measure. (For example, area of lune AA' is $\frac{a}{2\pi}$ times the total area of the sphere, which is $\frac{a}{2\pi} * 4\pi = 2a$.) If we add the area of each pair of lunes together, we will count the area of $\triangle ABC$ three times and the area of $\triangle A'B'C'$ three times. (*Explain why.*)

Of course, $\triangle ABC \cong \triangle A'B'C'$. When we use all of this information, we can say that the sum of the areas of the lunes is equal to the area of the sphere plus four times the area of $\triangle ABC$ (*Be sure you can explain why we add* four *times the area of $\triangle ABC$.*), giving the equation:

$$2(2\alpha) + 2(2\beta) + 2(2\gamma) = 4\pi + 4(\text{area of } \triangle ABC)$$

which simplifies to:

$$\alpha + \beta + \gamma - \pi = \text{area of } \triangle ABC$$

You can find an interactive version of this proof at a website written by an author and Dr. Stephen Szydlik - http://www.uwosh.edu/faculty_staff/szydliks/elliptic.shtml

Connections to Teaching:

I have never let my schooling interfere with my education.
Mark Twain

Why should students study non-Euclidean geometries? We think there are many reasons, the first of which is that one way we learn about what something *is* is by seeing what it *is not*; non-Euclidean geometry gives us a useful contrast to our standard high school geometry. It also brings into sharp focus the importance of *axioms* (one change in one axiom and you get a whole new geometry with different theorems) and mathematical *definitions* (for example think about what happened when we changed our definition of *distance* in the case of Taxi-cab geometry). Finally, scientists are gaining more and more evidence that our universe is not Euclidean space. Non Euclidean geometry is thus becoming increasingly important to an understanding of astronomy.

These geometries provide students opportunities to model and explore different types of spaces. In their paper in the journal *Mathematics Teaching in the Middle School*, Sharp and Heimer (2002) describe their experience having a sixth-grade class explore geometry on a sphere using beach balls in much the way you did in the class activity. Students defined what was meant by a *line* on a sphere, and explored lunes, triangles and other polygons. Finally, students applied what they'd learned to measurement on a globe. For example, sixth graders learned to use great circles (rather than lines of latitude) to find the shortest route between various cities on the planet. *What does typically happen when the globe is made into a flat map? What is distorted and in what way?*

Sharp and Heimer claimed that an experience with a non-Euclidean geometry helped their students to broaden their understandings of geometry. For example, children observed that parallel lines are impossible on a sphere and the authors argued that this sort of observation "... lays the foundation for the formation of informal deductions, a vital skill in geometric thinking, whether on the plane or the sphere" (p. 185). *What do they mean by this?*

Homework:

Do not worry about your difficulties in Mathematics. I can assure you mine are still greater.
Albert Einstein

1) Do all of the italicized things in *Read and Study* section.

2) Do all the italicized things in the *Connections* section.

3) In spherical geometry how many perpendicular lines can be drawn to a given line through a point not on that line? Does the answer to this question depend upon the location of the point in relation to the line? If so, describe the different cases and explain why you have more than one perpendicular to the line in some cases.

4) Does the Pythagorean Theorem hold in spherical geometry? If yes, support your answer with a proof. If no, support your answer with a counterexample.

5) Given that the angle sum of any spherical triangle is greater than 180°, make an argument (different from the one given in the reading) that rectangles do not exist in spherical geometry.

6) Determine which of the Euclidean triangle congruence theorems are true in spherical geometry. Support your answers with an argument or a counterexample.

7) Use the website at http://www.uwosh.edu/faculty_staff/szydliks/elliptic/elliptic.htm to explore similar triangles in spherical geometry. Can spherical triangles be similar but not congruent? Make an argument to support your answer. What does this say about the AAA Theorem?

8) Does the Isosceles Triangle Theorem hold for spherical triangles? Support your answer.

9) On the sphere draw a line you can consider the equator and let N be the point that would be the north pole. Mark two points, A and B, on the equator such that the measure of $\angle ANB$ is 90 degrees. Let C, D, and E be the midpoints of AB, AN, and BN (the minor segments), respectively.
        a) Explain and illustrate why CN, DB, and AE intersect in a common point, F.
        b) Find the angle sum of the spherical triangle ACF.

Class Activity 23: Life on a Hyperbolic World

Geometry is a skill of the eyes and the hands as well as of the mind.
Jean Pedersen

In this activity you will explore some of the properties of the geometry that results when "flatness" is replaced by "constant *negative* curvature." To do so we need a physical model to play with – and first you will need to make this model. Take the two sheets of regular heptagons from Appendix D, carefully cut out each heptagon and then tape the heptagons together at the edges, three to a vertex. Don't be surprised that they do not lie flat - recall that the vertex angle measure in a regular heptagon is about 128.57° and so three heptagons sum to more than 360°. Work with a partner and together make one sheet of hyperbolic paper to use in these explorations. (You will also need a length of string, a protractor, and colored markers or pencils.) Your final result should look like this:

1) What will a straight line look like on the hyperbolic plane? Use the same concept of straightness that we used on the sphere (the string method) and draw several straight lines on your hyperbolic paper. Do you think these lines are finite in length like those on the sphere – or are they infinite like lines on the flat Euclidean plane? (Remember there is nothing except time to keep you from adding more heptagons to all the edges of your hyperbolic paper. You are working with a piece of the hyperbolic plane, just like a regular 8 ½ x 11 sheet of paper is a piece of the Cartesian plane.)

(This activity is continued on the next page.)

2) Can you draw parallel lines on your hyperbolic paper? (Make certain you are using the string method to draw lines.) Can you make an argument that the lines you drew do not intersect somewhere on an extension of your paper? What do you notice about the distance between hyperbolic lines that do not intersect? How does this differ from Euclidean parallel lines?

3) Now choose a point on one of your parallel lines. Can you draw another line through that point that is also parallel to the first line? How many hyperbolic lines can be drawn parallel to the first line through this same point? (If your original pair of parallel lines are quite close together, it will be easier to answer this question if you choose a point farther away from one of the lines and see how many parallel lines you can draw through that point.)

4) There is a Euclidean theorem stating that two lines that are both parallel to the same line are also parallel to each other. Do you think this theorem holds in hyperbolic geometry? Why or why not?

5) Draw a small triangle and a large triangle (one that covers at least 1/4 of the paper) on your hyperbolic paper. Make certain that the sides of your triangles are actually straight line segments by using the string method to make the triangle. Determine a method to measure the angles of the triangles using your protractor and then measure each of the angles in both of the triangles. What is the angle sum of the small triangle? Of the large triangle? Make a conjecture about the angle sum of a hyperbolic triangle.

Read and Study:

Out of nothing I have created a strange new universe.
Janos Bolyai

The story of the development of hyperbolic geometry really begins with Euclid. Recall that he chose five postulates for his axiomatic system – the first four were generally accepted, but the fifth postulate (the Parallel Postulate) caused problems from the very beginning. First, it was lots more complicated than the others. Second, it did not seem as 'self-evident.'

Many mathematicians have tried to prove the Parallel Postulate from the other four, thinking it too complex a statement to accept without proof. In the 1700's, the Italian mathematician Saccheri mounted a concerted effort to show that if the Parallel Postulate was replaced by one that allowed more than one parallel, the resulting theorems would contradict themselves. While he found many interesting results, he did *not* find the contradiction he sought. However, he was so sure that the Parallel Postulate of Euclid was the only true case, he concluded his work by saying (without proof) that any other replacement postulate is absolutely false because it is "repugnant to the nature of the straight line."

A century later the famous German mathematician Gauss came to the conclusion that the 5th postulate is truly independent of the others. **In other words it cannot be proved using the other postulates (axioms) and nor does it contradict them**. **Furthermore, it can be replaced by alternative postulates which will yield interesting and consistent geometries different from Euclidean geometry.** *Read this paragraph again. It is important*.

However, Gauss was not willing to risk his significant mathematical reputation by publishing his results. And so it was left to two unknowns, Hungarian János Bolyai and Russian Nikolai Lobachevsky, to independently publish their findings of this strange new geometry we now call hyperbolic geometry.

As an axiom system, hyperbolic geometry retains all the axioms of Euclidean geometry except the Parallel Postulate, replacing it with the **Hyperbolic Parallel Postulate**: **Given a line and a point not on that line, there are at least two lines through that point parallel to the given line.** (In Spherical Geometry, Postulates I and III of Euclid are violated as well as the Parallel Postulate. *Explain how*.)

Of course, as we have already seen in our look at taxicab geometry, making just one change can result in a very different geometry. Hyperbolic geometry is no exception. Physically, we can understand the difference between Euclidean, hyperbolic, and spherical geometry by considering the curvature of the surface of a plane in each.

The Euclidean plane is flat; the spherical plane is curved positively so that it closes upon itself; and the hyperbolic plane is curved negatively so that standing at any one point the surface curves up along one direction and curves down along the perpendicular direction, like standing in the middle of a saddle or a Pringle potato chip. Alternatively, we can think about the sum of the interior angles in a triangle on each surface. If the sum is less than 180°, the surface is hyperbolic; if the sum equals 180°, the surface is Euclidean; and if the sum is greater than 180°, the surface is spherical. Below are pictures of surfaces with different curvatures. The hyperboloid has negative curvature, the cylinder has zero curvature, and the sphere has positive curvature.

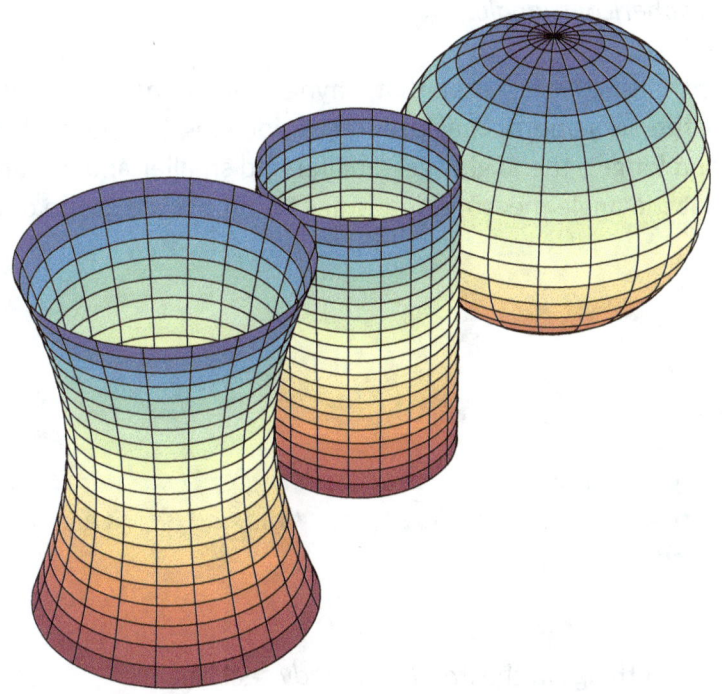

So how does this change in curvature (or equivalently, this change in the parallel postulate) change the geometry? Like Euclidean geometry, the hyperbolic plane is infinite and unbounded and so are hyperbolic lines. If we were to walk along a hyperbolic line in one direction, we would never return to our starting point, as we do in spherical geometry. We have an abundance of parallel lines, but, unlike Euclidean geometry, no two parallel lines are equidistant. There are actually two types of parallel lines. In one case, two parallel lines will be closest to each other at their single common perpendicular and then diverge from each other as you move away from that common perpendicular in either direction. In the other case, two parallel lines are asymptotic in one direction and divergent in the other. *Think about this*. If we have a pair of "lines" that are equidistant, one of the "lines" is not a line, but a curve. This is similar to the situation on the sphere where the equator and the 10° latitude marking are equidistant, but only the equator is a line.

We have triangles and other polygons in hyperbolic geometry, but, once again, they behave differently. There is only one hyperbolic line segment between two points so hyperbolic triangles are well-defined using the Euclidean definition. But the angle sum of a hyperbolic triangle is not constant and is always *less* than 180°. Furthermore, the area of a hyperbolic triangle gets larger as the angle sum gets smaller. And we can make the angle sum smaller by making the side lengths longer. (*Check out your results from the Class Activity. Do they support these claims?*)

As in spherical geometry, there is a formula for finding the area of a hyperbolic triangle that depends only on the measures of its angles: $A = \pi - (\alpha + \beta + \gamma)$. (*Notice the relationship with the area formula for a spherical triangle.*)

This formula shows us that the largest area that a hyperbolic triangle can have is p. As the angle sum approaches zero, the area approaches p, and the side lengths approach infinite length. So as our sides get longer and longer, the angles get smaller and smaller and our area never gets larger than p. This means that the angle measure determines not only the shape of the triangle but also its size.

Homework:

> *For God's sake, please give [hyperbolic geometry] up. Fear it no less than the sensual passion, because it too, may take up all your time and deprive you of your health, peace of mind and happiness in life.*
>
> *Wolfgang Bolyai (Janos' Father)*

1) Do all the italicized things in the *Read and Study* section.

2) Given that the angle sum of any hyperbolic triangle is less than 180°, argue that rectangles do not exist in hyperbolic geometry.

3) Do similar but not congruent triangles exist in hyperbolic geometry? What about the AAA Theorem? Support your answer.

4) Does the Isosceles Triangle Theorem hold for hyperbolic triangles? Support your answer.

5) There is a Euclidean theorem stating that two lines that are both parallel to the same line are also parallel to each other. Does this theorem hold in hyperbolic geometry? Support your answer.

6) Can we build a set of railroad tracks on a hyperbolic plane? Support your answer?

7) Can a right-angled regular pentagon exist on the hyperbolic plane? Support your answer.

Class Activity 24: Life in a Fractal World

The most exciting phrase to hear in science, the one that heralds new discoveries is not 'Eureka!' but 'That's funny...'

Isaac Asimov

In this activity you will create a famous fractal, the Koch Snowflake, and then investigate several of its properties. To create any fractal, we must apply a process to an initial geometric object and then apply the same process to the resulting object and then apply the same process to the resulting object and then apply the same process to the resulting object and then ... you get the idea.

We call such a procedure an **iterative process** and the object in each step is called an **iteration**. When the iterative process produces objects that are increasingly complex, but similar to the first iteration on a smaller and smaller scale, the 'final' iteration is a **fractal**. In theory, the process is repeated indefinitely, so there really is no final iteration but rather limiting object that is the actual fractal. Don't worry; we'll only produce three iterations of the Koch Snowflake.

To create the Koch Snowflake, take an equilateral triangle (the initial geometric object) and apply the following iterative process to *each side* of the triangle.

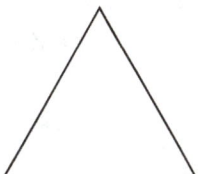

Step 1: Divide each line segment into thirds and remove (erase) the middle third.

Step 2: Replace the middle third with two sides of an equilateral triangle whose side length is the same as the length of the middle third you removed.

The following picture shows the process applied *once* to *one* side of the original triangle.

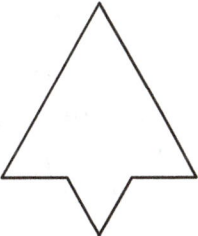

(This activity is continued on the next page)

1) Construct an equilateral triangle with sides approximately 2 inches long and use it to create the first three iterations of the Koch Snowflake. When you have finished you should have a separate drawing for each iteration. *Assume the side length of the original triangle is **one unit** in answering the following questions.*

2) What is the perimeter of the original triangle? The first iteration? The second iteration? The third iteration? Look for a pattern and make a conjecture for the perimeter of the nth iteration. What about the perimeter of the Koch Snowflake (the "infinite" iteration)?

3) What is the area of the original triangle? The first iteration? The second iteration? The third iteration? (It will be simpler to see a pattern if you use a non-standard unit for area – we suggest using the area of the smallest triangle in the third iteration as the unit.) Look for a pattern and make a conjecture for the area of the nth iteration. What about the area of the Koch Snowflake (the "infinite" iteration)?

Read and Study:

Big gaskets are made of little gaskets,
The bits into which we slice 'em.
And little gaskets are made of lesser gaskets
And so ad infinitum.

From http://classes.yale.edu/fractals/

Take a close look at the clouds, mountain ridges, lakeshores and icebergs in the two pictures below (taken by an author in Alaska). What geometric shape can be used to adequately describe the intricacies of their boundaries? A spherical triangle? A Euclidean circle? A hyperbolic polygon? No. Nothing we have studied thus far comes close to approximating the complexity of these natural shapes, particularly when they are examined on a small scale.

Benoit Mandelbrot is the mathematician credited with finding the geometric structure underlying these complicated natural shapes. In 1975 he coined the word *fractal* (from the Latin word *fractus*

meaning broken or fractured) to describe the convoluted curves and surfaces that can be used to model natural shapes. The key to his understanding was his observation that many real phenomena, such as coastlines, mountains and lungs, have a roughly *self-similar* shape: The smaller features of these objects have approximately the same shape and complexity as the larger features do. That is, a small portion of a mountain ridge will look approximately like an entire mountain ridge when magnified. *Think about this*.

Below is a computer-generated fractal picture (*not* a real picture) of ridges cut by a stream. It looks real doesn't it?

Mandelbrot used the concepts of self-similarity and complexity under magnification to describe certain mathematical sets that are fractal. A famous example, called the Mandelbrot set, has a boundary that is a mathematical fractal.

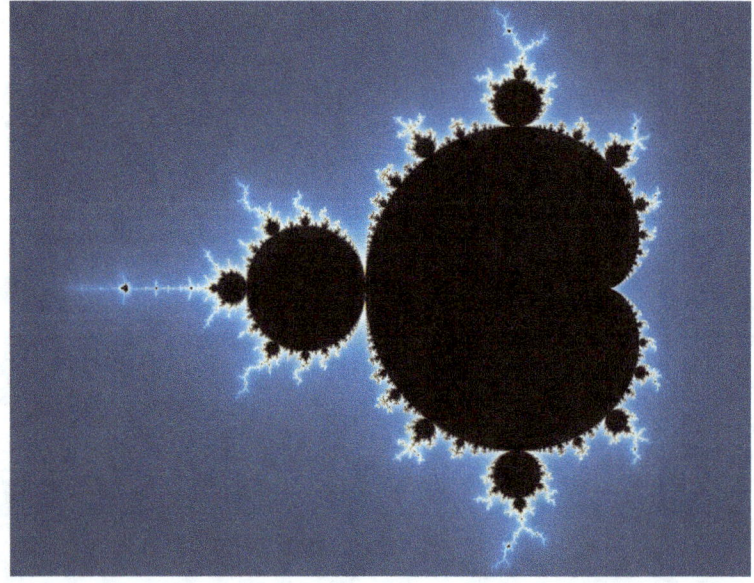

The Mandelbrot Set from http://en.wikipedia.org/wiki/Fractal

Approximate fractals are easily found in nature. These objects display self-similar structure over many magnifications. Examples include clouds, snowflakes, mountains, river networks, and broccoli. Trees and ferns are also fractal in nature and can be modeled on a computer by using a recursive (iterative) algorithm. This recursive nature is obvious in these examples — a branch from a tree or a frond from a fern is a miniature replica of the whole: not identical, but similar in nature.

Fractals provide a good model for many organs of the body, such as the lungs. The trachea splits into the bronchial tubes, which in turn split into shorter and narrower tubes. Even the embryonic development of the lung is an iterative process. The convoluted surface of the lung greatly increases its area while keeping its overall volume small. The large surface area is biologically essential because the amount of carbon dioxide and oxygen that the lungs can exchange is roughly proportional to their surface area. Using a light microscope, biologists found approximately 80 m² of surface area in a lung (roughly the floor space of a small house). The higher magnification of an electron microscope yielded approximately 140 m². Scientists have estimated the fractal dimension of a lung to be 2.17 (Thomas Q. Sibley. *The Geometric Viewpoint*. p. 220 – 221). We'll tell you what we mean by that in a minute.

All of these examples point out three necessary characteristics of a **fractal**:

1) it is self-similar (at least approximately);
2) it can be defined by an iterative process; and
3) it has a non-integer dimension that it larger than its geometric dimension.

(Note that not all self-similar objects are fractal. For example a line is self-similar, but its dimension is one, so it is not a fractal.)

Let's talk some more about *dimension* for a self-similar object. We will determine "dimension" by doubling its length and seeing how many copies of the original object we get. The dimension is the exponent to which you must raise the scaling factor (2 for doubling) in order to get the number of copies produced by that scaling.

A line segment has dimension one because when you double the length of the segment you get two copies of the segment and $2^1 = 2$. A square has dimension 2 because when you double the length of the side you get four copies of the square and $2^2 = 4$. A cube has dimension 3 because when you double the length of the edge you get eight copies of the cube and $2^3 = 8$. We can write this relationship as a formula as follows:

$$s^d = n \quad or \quad d = \frac{\log n}{\log s}$$

Here *s* is called the **scaling factor**, *d* is the **dimension**, and *n* is the number of copies produced. (*Try to explain the second version of the formula. How do we solve the first equation for d?*)

If we use this formula on the Koch Snowflake, we have *s* = 3, *n* = 4, and $d = \dfrac{\log 4}{\log 3} = 1.26$.

The Koch Snowflake has a dimension of 1.26. (*Weird, huh? Make sure you can explain why s* = 3 *and n* = 4.)

In some way, the dimension is a measure of the complexity of the fractal. The Koch Snowflake is more complex than a straight line, but not as complex as a square (including the interior).

In the class activity you explored the perimeter and area of the Koch Snowflake (named for the Swedish mathematician who first created it in 1904). Did you discover the amazing fact that this fractal has an *infinite* perimeter but a *finite* area? In other words, you can draw a circle around the entire fractal enclosing it within a finite area, but the *boundary* of the enclosed fractal is infinite in length.

In the homework problems you will determine the perimeter, area, and dimension of several other fractals. Be on the lookout to see if infinite perimeter and finite area just might be a property of all fractals.

Connections to Teaching:

Learning is not compulsory ... neither is survival.

W. Edwards Deming

Your future students will like doing geometry on a Möbius strip, a beach ball or a torus – but they will love fractals. Not only do fractals look cool, but the infinite aspect is fascinating to Middle-Schoolers. We will present just one fractal activity here; you can certainly find many, many more online.

For a sample, go visit the website https://fractalfoundation.org/resources/fractivities/ for an accessible discussion of fractals with lots of lesson plans for use in 3-12 grades.

The Sierpinski Triangle is another fractal that your students can investigate.

So what is the **Sierpinski Triangle**? Here's the idea. Begin with an equilateral triangle:

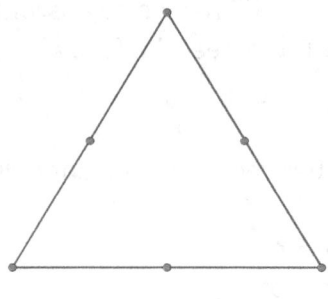

Step 0

Locate the midpoint of each side and create a new triangle by connecting those midpoints. Then remove that middle triangle.

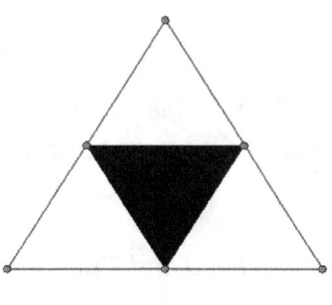

Step 1

Now do the same thing to each of the three resulting 'outside' triangles.

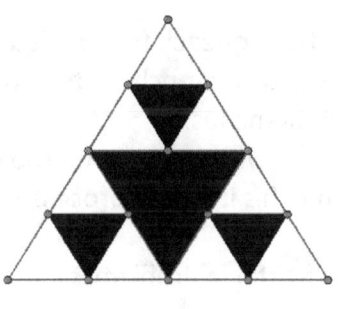

Step 2

Keep on going *forever*. (Recall that mathematical objects are *ideal* objects – so the idea of imaging what would happen if a process is repeated forever does not bother mathematicians.)

The resulting fractal is the Sierpinski Triangle. *Why is it a fractal? Use the definition to explain this. Suppose that the original triangle had an area of 1 u^2. Find a formula for the area at the nth step.*

A second activity involving the Sierpinski Triangle begins with triangle *ABC*. Pick any point inside the triangle as your starting point. The repeat the following process forever (or for as long as you wish):
- Roll a six-sided die.
- Mark a new point halfway between your current point and:
    Point A if it lands on 1 or 2.
    Point *B* if it lands on 3 or 4.
    Point *C* if it lands on 5 or 6.

An example of this game in action is here:
https://thewessens.net/ClassroomApps/Main/chaosgame.html

Homework:

> *...since geometry is the right foundation of all painting, I have decided to teach its rudiments and principles to all youngsters eager for art.*
>   Albrecht Durer, Course in the Art of Measurement

1) Do all the italicized things in the *Read and Study* section.

2) Do the problems in the *Connections* section.

3) Carefully sketch three iterations of each fractal idea.
   a) A stylized tree, where each branch splits into three others half as long. Begin with one trunk and three branches.
   b) A modified Koch curve, with a square on the middle third of a line segment, rather than a triangle. Apply this iterative process to each side of a square.

4) Find the perimeter of each fractal in Problem 3.

5) Find the limiting area of the fractal in Problem 3b.

6) Find the dimension of each fractal in Problem 3.

7) In the reading we discussed the concept of *self-similarity*. Another way to describe this property is to say that a self-similar object can be composed of smaller similar copies of itself. Which of the following geometric objects are self-similar: a line segment, a triangle, a

square, a trapezoid, a hexagon, a circle? Which of the self-similar objects are also fractals? Why?

8) Pictured below are the first four iterations of the box fractal. Write the instructions for the iterative process that creates it. What is the perimeter and area of the last iteration shown if the side of the original square is of length one? What is the dimension of the box fractal?

Summary of Big Ideas from Chapter Three

Man's mind, once stretched by a new idea, never regains its original dimensions.
Oliver Wendell Holmes

- We can change our geometry by changing the space, like the Klein bottle, sphere, or torus, the way we measure distance, like in taxi-cab geometry, or by adjusting an axiom, like in hyperbolic geometry.

- Taxi-cab geometry is a non-Euclidean geometry that middle grades students can explore.

- Several theorems from Euclidean geometry fail when applied to Spherical and Hyperbolic geometries.

- A Fractal is a geometric figure that is self-similar, that can be defined by an iterative process, and has a non-integer dimension.

Appendices

References:

- Adams, T. L. & Aslan-Tutak, F. (2005) Serving Up Sierpinkski! *Mathematics Teaching in the Middle School, 11 (5)*, p. 248-253.

- Battista, M. (2007). The development of geometric thinking. In the *Second Handbook of Research on Mathematics Teaching and Learning*, F. Lester (Ed.). NCTM: Information Age Publishing.

- *Common Core State Standards* as found in January 2012 at http://www.corestandards.org/assets/CCSSI_Math%20Standards.pdf

- *Mathematical Quotations Server* (MQS) at math.furman.edu.

- National Council of Teachers of Mathematics. (2006). *Curriculum Focal Points for Prekindergarten through Grade 8 Mathematics: A Quest for Coherence*. Reston, VA: NCTM.

- National Council of Teachers of Mathematics. (2000). *Principles and Standards for School Mathematics*. Reston, VA: NCTM.

- Poole, J. T. (2002). *Elements*. Found on January 10, 2012 at http://math.furman.edu/~jpoole/euclidselements/euclid.htm Department of Mathematics, Furman University, Greenville, SC.

- Sharp, J. & Heimer, C. (2002). What happens to geometry on a sphere? *Mathematics Teaching in the Middle School, 8 (4)*, p. 182.

- Shulman, L. S. (1985). On teaching problem solving and solving the problems of teaching. In E. A. Silver (Ed.), *Teaching and Learning Mathematical Problem Solving: multiple research perspectives* (pp. 439-450). Hillsdale, NJ: Erlbaum.

- Sibley, T. Q. (1997) *The Geometric Viewpoint: A Survey of Geometries*. Addison-Wesley.

Euclid's Postulates and Propositions

Euclid's *Elements*

This presentation of *Elements* is the work of J. T. Poole,
Department of Mathematics, Furman University, Greenville, SC.
© 2002 J. T. Poole. All rights reserved.

POSTULATES

Let the following be postulated:

1. To draw a straight line from any point to any point.

2. To produce a finite straight line continuously in a straight line.

3. To describe a circle with any center and distance.

4. That all right angles are equal to one another.

5. That, if a straight line falling on two straight lines make the interior angles on the same side less than two right angles, the two straight lines, if produced indefinitely, meet on that side on which are the angles less than the two right angles.

COMMON NOTIONS

1. Things which are equal to the same thing are also equal to one another.

2. If equals be added to equals, the wholes are equal.

3. If equals be subtracted from equals, the remainders are equal.

4. Things which coincide with one another are equal to one another.

5. The whole is greater than the part.

BOOK I   PROPOSITIONS

Proposition 1.
>On a given finite straight line to construct an equilateral triangle.

Proposition 2.
>To place at a given point (as an extremity) a straight line equal to a given straight line.

Proposition 3.
>Given two unequal straight lines, to cut off from the greater a straight line equal to the less.

Proposition 4.
>If two triangles have the two sides equal to two sides respectively, and have angles contained by the equal straight lines equal, they will also have the base equal to the base, the triangle will be equal to the triangle, and the remaining angles will be equal to the remaining angles respectively, namely those which the equal sides subtend.

Proposition 5.
>In isosceles triangles the angles at the base are equal to one another, and, if the equal straight lines be produced further, the angles under the base will be equal to one another.

Proposition 6.
>If in a triangle two angles be equal to one another, the sides which subtend the equal angles will also be equal to one another.

Proposition 7.
>Given two straight lines constructed on a straight line (from its extremities) and meeting in a point, there cannot be constructed on the same straight line (from its extremities), and on the same side of it, two other straight lines meeting in another point and equal to the former two respectively, namely each to that which has the same extremity with it.

Proposition 8.
>If two triangles have the two sides equal to two sides respectively, and have also the base equal to the base, they will also have the angles equal which are contained by the equal straight lines.

Proposition 9.
>To bisect a given rectilineal angle.

Proposition 10.
>To bisect a given finite straight line.

Proposition 11.
>To draw a straight line at right angles to a given straight line from a given point on it.

Proposition 12.
> *To a given infinite straight line, from a given point which is not on it, to draw a perpendicular straight line.*

Proposition 13.
> *If a straight line set up on a straight line make angles, it will make either two right angles or angles equal to two right angles.*

Proposition 14.
> *If with any straight line, and at a point on it, two straight lines not lying on the same side make the adjacent angles equal to two right angles, the two straight lines will be in a straight line with one another.*

Proposition 15.
> *If two straight lines cut one another, they make the vertical angles equal to one another.*

Proposition 16.
> *In any triangle, if one of the sides be produced, the exterior angle is greater than either of the interior and opposite angles.*

Proposition 17.
> *In a triangle two angles taken together in any manner are less than two right angles.*

Proposition 18.
> *In any triangle the greater side subtends the greater angle.*

Proposition 19.
> *In any triangle the greater angle is subtended by the greater side.*

Proposition 20.
> *In any triangle two sides taken together in any manner are greater than the remaining one.*

Proposition 21.
> *If on one of the sides of a triangle, from its extremities, there be constructed two straight lines meeting within the triangle, the straight lines so constructed will be less than the remaining two sides of the triangle, but will contain a greater angle.*

Proposition 22.
> *Out of three straight lines, which are equal to three given straight lines, to construct a triangle: thus it is necessary that two of the straight lines taken together in any manner should be greater than the remaining one.* [I.20]

Proposition 23.
> *On a given straight line and at a point on it to construct a rectilineal angle equal to a given rectilineal angle.*

Proposition 24.
>If two triangles have the two sides equal to two sides respectively, but have the one of the angles contained by the equal straight lines greater than the other, they will also have the base greater than the base.

Proposition 25.
>If two triangles have the two sides equal to two sides respectively, but have the base greater than the base, they will also have the one of the angles contained by the equal straight lines greater that the other.

Proposition 26.
>If two triangles have the two angles equal to two angles respectively, and one side equal to one side, namely, either the side adjoining the equal angles, of that subtending one of the equal angles, they will also have the remaining sides equal to the remaining sides and the remaining angle to the remaining angle.

Proposition 27.
>If a straight line falling on two straight lines make the alternate angles equal to one another, the straight lines will be parallel to one another.

Proposition 28.
>If a straight line falling on two straight lines make the exterior angle equal to the interior and opposite angle on the same side, or the interior angles on the same side equal to two right angles, the straight lines will be parallel to one another.

Proposition 29.
>A straight line falling on parallel straight lines makes the alternate angles equal to one another, the exterior angle equal to the interior and opposite angle, and the interior angles on the same side equal to two right angles.

Proposition 30.
>Straight lines parallel to the same straight line are also parallel to one another.

Proposition 31.
>Through a given point to draw a straight line parallel to a given straight line.

Proposition 32.
>In any triangle, if one of the sides be produced, the exterior angle is equal to the two interior and opposite angles, and the three interior angles of the triangle are equal to two right angles.

Proposition 33.
>The straight lines joining equal and parallel straight lines (at the extremities which are) in the same directions (respectively) are themselves also equal and parallel.

[Proposition 34]().
> *In parallelogrammic areas the opposite sides and angles are equal to one another, and the diameter bisects the areas.*

[Proposition 35]().
> *Parallelograms which are on the same base and in the same parallels are equal to one another.*

[Proposition 36]().
> *Parallelograms which are on equal bases and in the same parallels are equal to one another.*

[Proposition 37]().
> *Triangles which are on the same base and in the same parallels are equal to one another.*

[Proposition 38]().
> *Triangles which are on equal bases and in the same parallels are equal to one another.*

[Proposition 39]().
> *Equal triangles which are on the same base and on the same side are also in the same parallels.*

[Proposition 40]().
> *Equal triangles which are on equal bases and on the same side are also in the same parallels.*

[Proposition 41]().
> *If a parallelogram have the same base with a triangle and be in the same parallels, the parallelogram is double of the triangle.*

[Proposition 42]().
> *To construct, in a given rectilineal angle, a parallelogram equal to a given triangle.*

[Proposition 43]().
> *In any parallelogram the complements of the parallelograms about the diameter are equal to one another.*

[Proposition 44]().
> *To a given straight line to apply, in a given rectilineal angle, a parallelogram equal to a given triangle.*

[Proposition 45]().
> *To construct, in a given rectilineal angle, a parallelogram equal to a given rectilineal figure.*

[Proposition 46]().
> *On a given straight line to describe a square.*

Proposition 47.
> In right-angled triangles the square on the side subtending the right angle is equal to the squares on the sides containing the right angle.

Proposition 48.
> If in a triangle the square on one of the sides be equal to the squares on the remaining two sides of the triangle, the angle contained by the remaining two sides of the triangle is right.

Glossary

Acute angle – an angle with measure less than the measure of a right angle

Acute triangle – a triangle with three acute angles

Adjacent angles – two non-overlapping angles that share a vertex and a common ray

Affine plane – a geometry with parallel lines based on the affine set of axioms

Algorithm – a set of steps used to carry out a procedure

Alternate exterior angles – two angles (formed by a transversal of a pair of lines) that lie outside the lines and on opposite sides of the transversal

Alternate interior angles – two angles (formed by a transversal of a pair of lines) that lie between the lines and on opposite sides of the transversal

Altitude (of a triangle) – the line through a vertex that is perpendicular to the opposite side

Altitude (of a pyramid) – the line segment from the apex perpendicular to the base of the pyramid; also called the height

Altitude (of a prism) – a line segment perpendicular to the bases of the prism; also informally called the "height"

Analytic geometry – the use of a coordinate system to translate geometric problems into algebraic problems

Angle – the figure formed by two rays with a common endpoint

Angle bisector – the line through the vertex of an angle that divides the angle into two congruent angles

Antipodal points – points that are the endpoints of a diameter of a sphere

Apex (of a pyramid) – the common point of the non-base faces of a pyramid

Apex (of a cone) – the common point of the line segments that create a cone

Arc – the set of points on a circle between two given points of the circle (There are actually two arcs between any two given points; the shorter one is called the *minor* arc and the longer one is called the *major* arc.)

Area – the quantity of two-dimensional space enclosed by a plane figure

Attribute – a property of a geometric object that can be measured (such as length) or categorized (such as color)

Axiom – a statement that is true by assumption

Axiomatic system – a set of undefined terms, definitions, axioms, and theorems that create a mathematical structure

Axis (of a cone) – the line joining the apex to the center of the (circle) base

Axis of symmetry – a line in space around which a three-dimensional object is rotated

Base angles (of an isosceles triangle) – the angles that are opposite the congruent sides of an isosceles triangle

Bilateral symmetry – an object has bilateral symmetry when it has exactly one line of reflectional symmetry

Bisect – to divide a geometric object (such as a line segment or an angle) into two congruent pieces

Boundary – the set of points that separate the inside of a closed planar object from the outside

Center (of a circle) – the point that is equidistant from all points on the circle

Central angle – an angle whose vertex is a center of a geometric object

Centroid – the point of intersection of the three medians of a triangle; also known to be the center of mass of the triangle

Chord – a line segment whose endpoints are distinct points on a given circle

Circle – the set of points that are the same distance from a given point, called the center

Circumcenter – the point of intersection of the three perpendicular bisectors of a triangle; also the center of the circle that circumscribes the triangle

Circumscribed circle – the circle that contains all the vertices of a polygon

Closed curve – a curve that starts and stops at the same point

Closure (of a set under an operation) – the property that the result of the operation on any two elements of the set is also an element of the set

Collinear points – points that lie on the same line

Complementary angles – two angles whose measures sum to the measure of one right angle

Composition of rigid motions – the combined actions of two rigid motions with the second motion applied to the image of the first motion

Concave polygon – a polygon for which at least one diagonal lies outside the polygon

Concurrent lines – three or more lines that intersect in the same point

Cone (circular) - a three-dimensional geometric object consisting of all line segments joining a single point (called the apex) to every point of a circle (called the base)

Congruent objects – two geometric objects are congruent if one object is the image of the other under a rigid motion of the plane.

Conic sections – the four curves (circle ellipse, hyperbola, and parabola) formed when a plane intersects a double cone.

Conjecture – a guess or a hypothesis about what is true in an axiom system.

Converse (of "If A, then B.") – "If B, then A," where A and B are statements

Convex polygon – a polygon all of whose diagonals lie inside the polygon

Consistent (set of axioms) – one in which it is impossible to deduce from these axioms a theorem that contradicts any axiom or previously proved theorem

Construction – creating a geometric object using only straight line segments and circles (Euclid's first, second, and third axioms)

Contrapositive (of "If A, then B.") – "If not B, then not A," where A and B are statements

Coordinate (Cartesian) plane – a model of Euclidean geometry in which each point is identified by two coordinates, the first of which represents the horizontal distance of the point from the y-axis and the second of which represents vertical distance from the x-axis. (The x- and y-axes are perpendicular and lie in the same plane.)

Coplanar lines – lines that lie in the same plane

Corresponding angles - two angles (formed by a transversal of a pair of lines) that lie on the same side of the transversal and also lie on the same side of the pair of lines

Corresponding points – a pair of points, one of which is the original point and the other of which is the image of that point under a rigid motion

Counterexample – an example that shows a conjecture is false

Curve – a set of points drawn with a single continuous motion

Cylinder (circular) – a three-dimensional geometric object consisting of two parallel and congruent circles (and their interiors) and the parallel line segments that join corresponding points on the circles

Deductive reasoning – the process of using logic to arrive at a conclusion

Definition – a statement of the meaning of a term, word, or phrase

Degree – a unit of angle measure for which a full turn about a point equals 360 degrees

Diagonal – the line segment joining two non-adjacent vertices of a polygon

Diameter – a line segment through the center of a circle whose endpoints lie on the circle

Dilation: (with center P and scale factor $k > 0$) a motion of the plane in which the image of P is P and the image A' of any other point A is on the ray PA so that the distance PA' is k times the distance PA.

Dimension (of a real space) – the number of mutually perpendicular directions needed to describe the location of the set of points in that space

Edge – the line segment (side) that is shared by two faces of a polyhedron

Ellipse – the set of points P in the plane such that the sum of the distances from P to two given points F_1 and F_2 is constant. The points F_1 and F_2 are called the foci of the ellipse.

Equiangular (polygon) – a polygon all of whose vertex angles are congruent

Equilateral (polygon) – a polygon all of whose sides are congruent

Euclidean model – a model of the geometry of the infinite flat plane based on the axiom system first established by Euclid

Euler's line – the line containing the circumcenter, the centroid, and the orthocenter of a triangle

Exterior angle – the angle formed by a side of a polygon and the extension of an adjacent side

Face – a polygon (with interior) that forms a portion of the two-dimensional surface of a polyhedron

Finite geometry – a geometry that consists of a finite number of points and their relationships

Fixed point – a point P whose image under a rigid motion is P

Fractal – an object that results from applying an iterative process in which each iteration is increasingly complex, but self-similar

Function – a rule that assigns to each element of a set S an element of set T in such a way that every element in S is paired with an element of T and no element of S is assigned to more than one element of T

Glide reflection – a rigid motion that is the composition of a translation and a reflection in which the line of reflection and the translation vector are parallel

Great circle – the intersection of a sphere and a plane that contains the center of the sphere

Height (of a triangle) – length of the line segment from a vertex perpendicular to the opposite side

Hyperbola – the set of points P in the plane such that the difference of the distances from P to two given points F_1 and F_2 is constant

Hypotenuse – the side of a right triangle opposite the right angle

Identification space – a two dimensional model of an object that lives in higher dimensions. The model shows how sides are identified ("glued together")

Image (of a rigid motion) – the set of points that result from the motion of an object by a rigid motion of the plane

Incenter – the point of intersection of the three angle bisectors of a triangle; also the center of the inscribed circle

Incircle (inscribed circle) – the circle that is tangent to all sides of a polygon

Inductive reasoning – the informal process of using examples to come to a conclusion, create a conjecture, or to generalize

Inscribed circle – the circle that is tangent to each side of a polygon

Inscribed angle (of a circle) – an angle formed by two chords with a common endpoint

Intersection (of two lines) – the point(s) the lines have in common

Intersection (of two sets) – the set of elements that are common to both sets

Isometry (of a set S) – a transformation of the set S that preserves distances, meaning that the distance between any points P, Q is equal to the distance between their images P', Q'

Isosceles – having at least one pair of congruent sides

Iterative process – an algorithm applied to an object and then to the result and then to the result and so forth. The object in each step of the process is called an iteration

Justification – an argument based on axioms, definitions, and previously proven results to show that a conjecture is true

Leg – a side of a right triangle opposite an acute angle

Length – the measure of a 1-dimensional object

Line – an undefined one-dimensional set of points understood to follow the shortest path (between every pair of points on the line) and to extend in opposite directions indefinitely

Line of reflection – the line about which an object is reflected to form its mirror image

Line segment – the set of points on a line between two given points, called the endpoints

Locus (definition) – a definition that describes a curve as a set of points in the plane

Logically equivalent (statements) – statements that have the same truth value in every case

Lune – a concave plane region bounded by two arcs of different radii

Major segment (of a great circle) – the larger of the two arcs determined by two distinct points on a great circle

Measure – to determine the quantity of an attribute (or of a fundamental concept such as time) using a given unit

Median – the line segment joining a vertex of a triangle to the midpoint of the opposite side

Midpoint – the point on a line segment that divides it into two congruent line segments

Minor segment (of a great circle) – the smaller of the two arcs determined by two distinct points on a great circle

Model – a representation of an axiom system in which each undefined term is given a concrete interpretation which allow the axioms to make sense

Net – a two-dimensional model that can be folded into a three-dimensional object

Obtuse angle – an angle with measure greater than the measure of a right angle

Obtuse triangle – a triangle with one obtuse angle

One-to-one (function) – a function from a set S to a set T in which no element of T is assigned to more than one element of S

Onto (function) – a function from a set S to a set T in which every element of T is assigned to some element from S

Order (of an affine plane) – the number of points on each line of the plane

Order (of a projective plane) – the number of points on each line of the plane less one

Order (of a rotational symmetry) – the number of different rotations that are a symmetry of an object

Orientation – the direction, clockwise or counterclockwise, of the reading of the vertices of a polygon in alphabetical order

Orthocenter – the point of intersection of the three altitudes of a triangle

Orthogonal (circles) – intersecting circles whose respective radii (or respective tangents) are perpendicular at the points of intersection

Parabola – the set of points *P* in the plane such that the distance from *P* to a given point *F* is equal to the distance from *P* to a given line *m*. Point *F* is called the focus of the parabola and line *m* is the directrix

Parallel lines – coplanar lines with no points in common

Parallelogram – a quadrilateral in which both pairs of opposite sides are parallel

Partition – a division of a geometric object into a set of non-overlapping objects whose union is the original object

Perimeter (of a plane object) – the length of the boundary of the object

Perpendicular bisector – the line through the midpoint of a line segment that is also perpendicular to the line segment

Perpendicular lines – two lines that intersect to form four right angles

Pi – the ratio of the circumference of a circle to its diameter; this ratio is an irrational number that is constant for all size circles and is approximately equal to 3.1415926

Planar curve – a curve that lies entirely within a plane

Plane – an undefined two-dimensional set of points understood to extend in all directions indefinitely

Plane of symmetry – a plane in space about which a three-dimensional object is reflected

Point – an undefined zero-dimensional object understood to be a location with no size

Polygon – a set of line segments that form a simple closed planar curve

Polyhedron (plural: polyhedra) – a finite set of polygons joined pair-wise along the sides of the polygons to enclose a finite region of space within one chamber

Postulate – an axiom

Prism – a polyhedron in which two of the faces are parallel and congruent (called the bases) and each of the remaining faces is a parallelogram that shares one side with each base.

Projective plane – a geometry in which there are no parallel lines based on the projective set of axioms

Proof – a deductive argument, written in formal mathematical language, that establishes the truth of a claim using axioms, definitions, and previously proven theorems

Pyramid – a polyhedron in which all but one of the faces is triangles that share a common vertex (called the apex); the remaining face may be any polygon and is called the base

Quadrilateral – a polygon with exactly four sides

Quantifier (in logic) – a word or phrase (such as "all" or "at least one") that indicates the size of the set to which the statement applies

Radius (plural: radii) – the line segment joining a point on a circle to the center of the circle

Ray – the set of points on a line beginning at a given point (called the endpoint) and extending in one direction on the line from that point

Rectangle – a quadrilateral with four right angles

Rectilinear angle – an angle formed by straight lines (as opposed to curves), nowadays simply referred to by *angle*

Redundant (set of axioms) – a set of axioms in which it is possible to prove at least one of the axioms from the other axioms

Reflection (in a line *l*) – a rigid motion of the plane in which the image of a point P on *l* is P, and if $A \ne P$ and if the image of A is A', then *l* is the perpendicular bisector of $\overline{AA'}$.

Reflectional symmetry (2-dimensional) – a reflection in which an object is divided by the line of reflection into two parts that are mirror images of each other

Reflectional symmetry (3-dimensional) – a reflection in which an object is divided by the plane of reflection into two parts that are mirror images of each other

Regular polygon – a polygon with all sides congruent and all vertex angles congruent

Regular polyhedron – a polyhedron whose faces are all the same regular polygon with the same number of faces meeting at each vertex

Rhombus (plural: rhombi) – a quadrilateral with four congruent sides

Right angle – an angle that forms exactly one fourth of a complete turn about a point

Right triangle – a triangle with one right angle

Rigid Motion of the plane (see also isometry) – a motion of the plane that preserves the distances between points

Rotation (about a point P through an angle q) – a rigid motion of the plane in which the image of P is P and, if the image of A is A', then $\overline{PA} \cong \overline{PA'}$ and $m\angle APA' = q$. Point P is called the center of the rotation

Rotational symmetry (2-dimensional) – a rotation about a point in which the image coincides with the original object

Rotational symmetry (3-dimensional) – a rotation about an axis of symmetry in which the image coincides with the original object

Scalene triangle – a triangle none of whose sides are congruent

Scaling – a transformation of the plane that causes either a magnification or a shrinking of an object in which the image remains similar to the original object

Scaling factor – the factor by which an object is magnified or contracted in a scaling

Secant – a line that intersects a circle in two distinct points

Sector – the portion of a circle and its interior between two radii

Shearing – a transformation of the plane that changes the shape of an object

Side – one of the line segments that make up a polygon

Similar (polygons) – polygons whose corresponding vertex angles are congruent and whose corresponding sides are proportional

Simple curve – a curve that does not intersect itself

Slope (of a line on the Cartesian plane) – the tangent of the angle of inclination the line makes with the positive x-axis

Space – an undefined term that denotes the set of points that extends indefinitely in three dimensions

Sphere – the set of points in (three-dimensional) space that are equidistant from a given point, called the center

Square – a quadrilateral with four right angles and four congruent sides

Supplementary angles – two angles whose measures sum to the measure of two right angles

Surface – the set of points that form the boundary of a solid three-dimensional object

Surface area – the sum of the areas of the faces of a closed 3-dimensional object

Symmetry (of an object) – a rigid motion of the object in which the image coincides with the original

Tangent (to a circle) – a line that intersects a circle in exactly one point

Taxicab geometry – a geometry of the infinite flat plane in which distance between points is measured as the sum of the vertical and horizontal distances between the two points

Theorem – a mathematical statement that is proven true

Transformation (of a set S) – a function from S to S that is both one-to-one and onto

Translation (by a vector RS) – a motion of the plane so that if A is any point in the plane and we call A' the image of A, then vector AA' and vector RS have the same length and direction

Transversal – a line which intersects two or more lines (each at a different point)

Trapezoid – a quadrilateral with exactly one pair of parallel sides

Triangle – a polygon with exactly three sides

Trivial rotation – the rotation of 360°; it is a rotational symmetry of every object

Undefined term – a term which has an intuitive meaning, but no formal definition

Union (of sets) – the set containing every element of each set

Vertex (plural: vertices) – the common endpoint of two adjacent sides of a polygon

Vertex angle – the angle formed by adjacent sides of a polygon

Vertex (of a polyhedron) – the intersection of two or more edges of a polyhedron

Vertical angles – a nonadjacent pair of angles formed by two intersecting lines

Volume – a measure of the capacity of a 3-dimensional object or, alternatively, the quantity of space enclosed by a 3-dimensional object

Polygon Cut-Outs (for Read and Study 10)

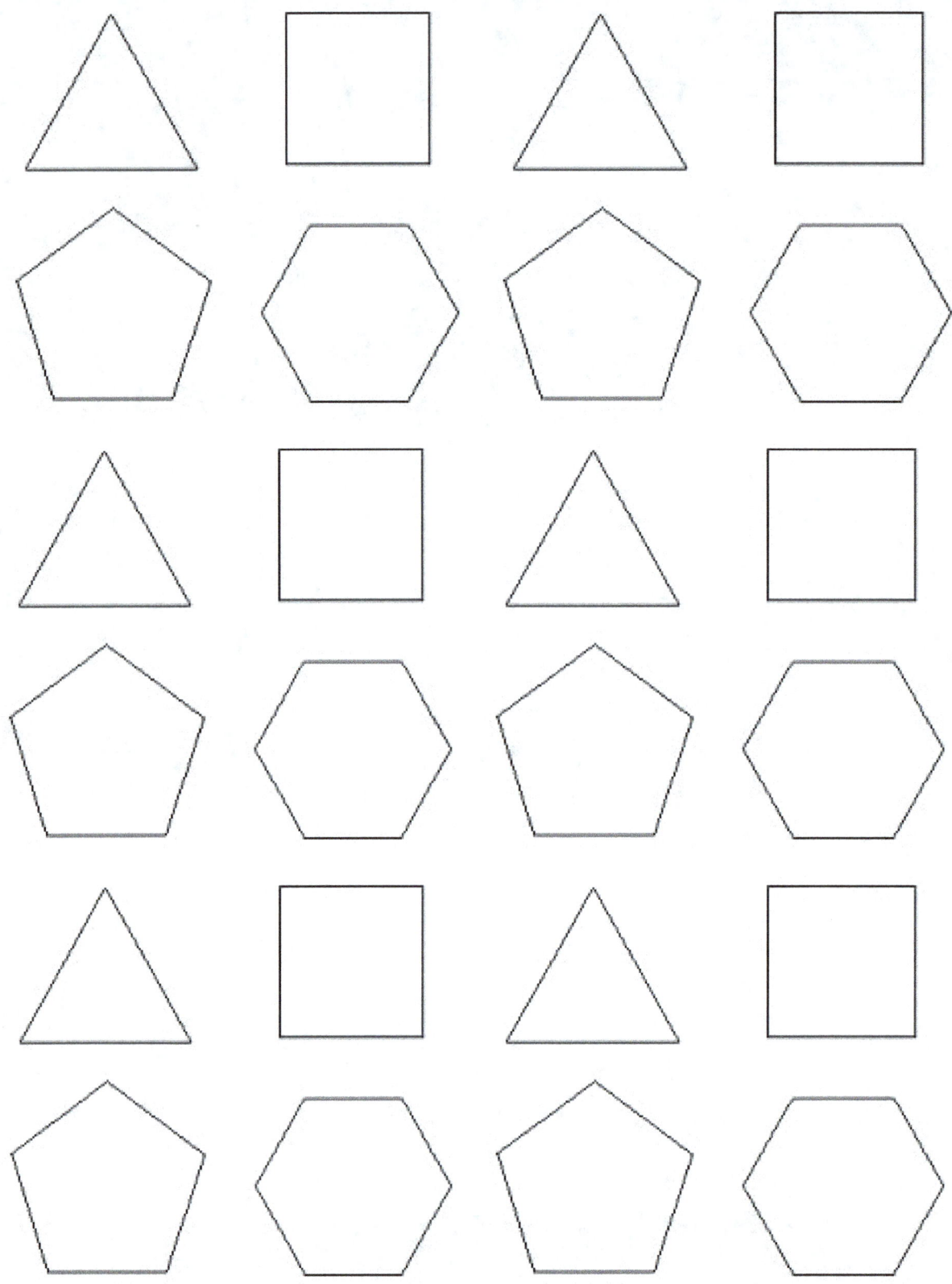

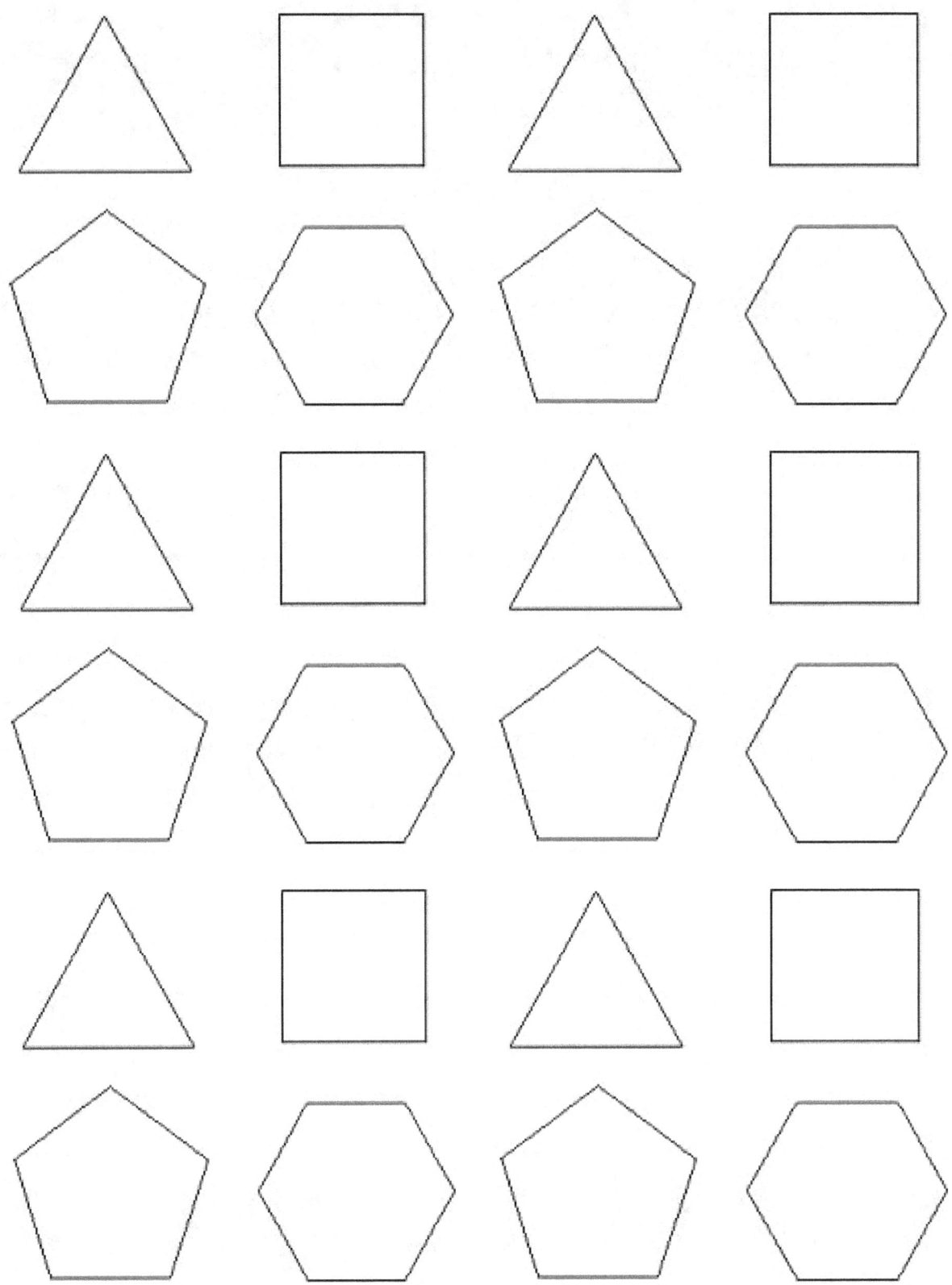

Hyperbolic Paper Template